METEOROLOGY
AT SEA

METEOROLOGY AT SEA

Ray Sanderson

STANFORD MARITIME
LONDON

Stanford Maritime Limited
Member Company of the George Philip Group
12-14 Long Acre London WC2E 9LP
Editor Phoebe Mason

First published in Great Britain 1982
Copyright © Ray Sanderson 1982

Set in Times Roman 10/11 and
Printed in Great Britain by
Ebenezer Baylis & Son Ltd
The Trinity Press, Worcester, and London
Drawings by Naomi Games

British Library Cataloguing in Publication Data

Sanderson, Ray
    Meteorology at Sea.
    1. Meteorology, Maritime
    I. Title
    551.5′0246238   QC994

ISBN 0 540 07405 5

Contents

Introduction

This book has grown from a series of lectures given by the author to a RYA/DoT Yachtmaster Certificate course at Greenwich. Therefore it is primarily aimed at yachtsmen, though it will also be found useful by those who go to sea in larger vessels.

Pure theory has been reduced to a minimum, especially with regard to conditions in the upper air since yachtsmen are more concerned with near surface conditions, primarily those accounting for the wind. Here the treatment of the forces involved is no more complicated than the resolution of everyday navigational problems. In any case the sailing yachtsman has a need to know just as much about his chief means of propulsion, the wind, as he does about his sails and auxiliary engine. The wind is also important when very strong, since together with poor visibility it provides the most serious hazards other than navigational ones (which are more predictable) that small vessels are likely to meet.

For these reasons the ways of the wind are given more detailed treatment than any other aspect of meteorology. In fact one section of the book is devoted to wind and visibility, the two potential hazards perhaps more feared than any other by yachtsmen at sea.

The book may also be found useful by those with a general interest in meteorology and by students taking A Level Geography in the General Certificate of Education.

Acknowledgements

I am grateful to the Director-General of the Meteorological Office for permission to publish this book, to Her Majesty's Stationery Office for permission to publish several photographs and the climatological tables, to the Hydrographic Department of the Navy for permission to use a part of the chapter on sea ice from NP100 (originally my own work), and to NOAA for most of the satellite pictures. My thanks, too, to Brian Wales-Smith for his constructive criticism of the text, to Phyllis Burton for typing from my manuscript, to Nichos Solomonides for plotting the surface and upper air charts, and last, but not least, to Jackie, Fiona and John who have had to put up with so much for so long.

Dedication

To the Blackpoll Warbler, a small land-bird which, having awaited the passage of a cold front over the New England coast, carries the post-front northwesterly winds to the region of Bermuda before turning southwestwards to use the prevailing northeasterlies for the remainder of its two thousand mile non-stop autumn migration to South America.

Diagrams

Tables

CHAPTER 1

The Atmosphere and its General Circulation

THE atmosphere is a layer of air which envelops the globe. Though it extends several hundred kilometres above the surface of the earth, it is relatively very shallow when compared with the earth's radius; it has often been referred to as 'the skin on a large apple'. This envelope of air, due to its own weight, is most dense at the surface of the earth becoming more rarefied with increasing height. The weight of the column of air vertically above a given place exerts a pressure which is called the Atmospheric Pressure (see Appendix 1).

There are various layers within the atmosphere which are differentiated by their temperature structures. Diagram 1 shows these layers and the vertical temperature structure throughout the 'whole' atmosphere. We shall be chiefly concerned in this book with the Troposphere, the lowest layer of the atmosphere, in which occurs all the earth's weather (cloud, rain, snow etc) and which contains three-quarters of the air within the atmosphere. The upper boundary of the troposphere is the Tropopause, where a temperature minimum occurs. The tropopause is important in that it provides an upper limit to weather and also to vertical movement of the air within the troposphere.

The layer above the tropopause ranges from about 11 up to 50 km where there is a secondary temperature maximum. The atmosphere is almost transparent to incoming radiation from the sun, most of which reaches the earth's surface resulting in the layer of air in contact with the ground being warmed; it is here that the highest temperatures in the troposphere normally occur. However, the greater part of the ultraviolet radiation from the sun, which would otherwise be extremely harmful to life on earth, is absorbed by ozone which occupies a layer of the atmosphere at about the Stratopause. This absorption gives rise to warming, hence the secondary temperature maximum at the stratopause. The tropopause can be thought of as a level with a minimum temperature between two layers with maxima. The Thermosphere is, again, warmed by the direct action of solar radiation accounting for yet another warm region in the atmosphere. The minimum between the thermosphere and stratopause maxima is at the Mesopause which lies at about 90 km. The whole layer above the stratopause is the Ionosphere, the layer which reflects medium and long wave radio transmissions and where the aurora ranges, at least in high latitudes.

The atmosphere is a fluid which is in

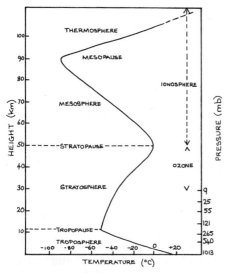

1. The vertical structure of the atmosphere

air within the troposphere, and in particular to the existence and movement of anticyclones and depressions, atmospheric pressure varies from place to place and from time to time. This variation, which leads to wind and weather, is due to two main reasons: differential heating from the sun, and the rotation of the earth.

Diagram 2 shows the effect of differential heating over the globe. The amount of heat radiation received from the sun at a given place depends on the season. We shall consider an average state for the whole year (this corresponds to the spring/autumnal equinoxes when the radiation received in each hemisphere is equal). Because the sun is overhead at the equator more heat radiation is received in this region than elsewhere. This leads to high temperatures in the equatorial belt. Since little radiation is received at both poles, temperatures are low there. This is the average generalized condition over the globe and is the primary driving force for the general circulation of the atmosphere around it. The hot air in the equatorial surface region expands and begins to rise vertically. This process continues until the rising air reaches the tropopause where it is

constant motion. Within this moving fluid eddies form and decay, the larger ones perhaps lasting up to several days or even a week or two. They are areas with a pressure minimum at their centre and are known as Depressions or Lows. There are other circulations within the troposphere where there is a pressure maximum at the centre; these are known as Anticyclones or Highs. Due to the constant movement of the

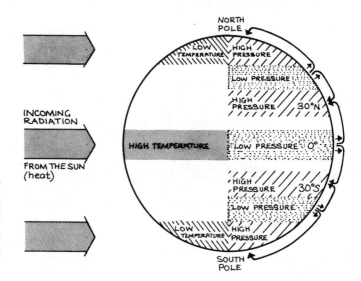

2. The global distribution of temperature and pressure. As a convenience, temperature is shown on the 'western' hemisphere and pressure over the 'eastern'. Both, of course, extend around the whole world.

deflected towards both poles.

It is now that the second factor in accounting for variations in atmospheric pressure becomes important. The rotation of the earth causes moving air to be deflected towards the right in the northern hemisphere and towards the left in the south. (The deflecting force is called the Geostrophic Force and will be explained later in Chapter 4 on winds.) The air at upper levels moving polewards is thus deflected to run eastwards. This deflection occurs in about latitudes 30°N and S. The air now running eastwards slowly descends to the earth's surface in these same latitudes. Since air (from the equatorial regions) is now being added into the regions around 30°N and S this results in the establishment of high pressure areas in these latitudes and between them a belt of lower pressure in the equatorial region. The high pressure areas are known as the Sub-Tropical Highs (the horse latitude highs of square-rig days); the low pressure between them is simply referred to as the Equatorial Low Pressure Belt (or Equatorial Low).

The coldness of the polar areas results in high pressure over both poles. This is accounted for by the fact that cold air contracts and becomes more dense and therefore heavier, resulting in a rise of pressure when measured at the surface of the earth. Between the polar and sub-tropical high pressure belts in each hemisphere lies a region of lower pressure centred in roughly 50° to 60°N and S; these are known as the Temperate Latitude Low Pressure Belts. Air rises throughout the troposphere in low pressure areas. The ascending air, then, in the temperate latitude low pressure belt turns poleward or equatorward at the tropopause, the former stream descending in high latitudes to enhance the polar highs and the latter finally adding to the sub-tropical highs. Thus the general distribution of atmospheric pressure over the globe, shown in Diagram 2, is of high pressure at both poles and over the sub-tropical regions in each hemisphere, and belts of lower pressure in about 50° to 60°N and S and in the equatorial region.

Nature constantly attempts to equalize the pressure over the whole globe; thus air begins to move from the high pressure regions towards the low pressure regions but it is deflected due to the earth's rotation to the right or left respectively in the northern and southern hemispheres. Diagram 3 shows the resulting average surface

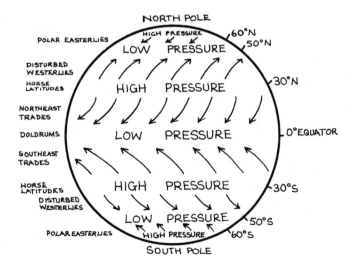

3. The main wind belts around the globe

wind pattern over the globe. It should constantly be referred to throughout the following description. Air from the polar high pressure regions moves, initially, equatorwards but it is soon deflected to run towards the west (easterly wind) in both hemispheres. Air from the sub-tropical highs, initially moving polewards on the one side and equatorwards on the other is, of course, also deflected. On the northern side of the sub-tropical high in the northern hemisphere the air is deflected to run towards the east, while on its southern side, the air is deflected to run towards the west. Due to the opposite deflecting effect in the southern hemisphere air also runs westwards in the region between the southern sub-tropical high and the equator, eastwards on the southern side of this high and westwards around the polar high over the Antarctic region, as shown in Diagram 3 which also gives the names of the various wind belts.

The regions between 50° and 60° latitude in each hemisphere are especially interesting; these are the regions where the polar easterlies run alongside the westerlies. As we shall see later, this is the region where travelling depressions are formed. (It can be seen that these adjacent wind directions are already appropriate to the direction of rotation in travelling depressions: anticlockwise in the northern hemisphere and clockwise in the south.)

Another interesting region is that between the sub-tropical highs. The 'easterly' winds here are known as the Northeast and Southeast Trades. They are much more persistent than in any other wind belt. They in turn, are sometimes divided by a tongue of Equatorial Westerlies.

All of the surface wind belts, easterly or westerly, have components either towards the poles or the equator. The 'net' flow of air is towards the equator at low levels whereas the net flow at high levels is towards the poles, though as Diagrams 2 and 3 show this is not by any means a simple circulation. This complex migration of air from ascent in the equatorial regions through to the trade winds replenishing the initial loss in this area is known as the General Circulation of the atmosphere.

We have seen that the main driving forces behind this circulation are differential heating and the deflecting force due to the rotation of the earth, resulting in the sub-tropical and polar anticyclones. These features are often referred to as the semi-permanent high pressure regions. Though they most frequently occupy the positions shown in Diagram 2, they are by no means permanent in these latitudes, since the sub-tropical high may be found several degrees north or south of its average position and the polar highs may be displaced to one side or the other of the pole.

However, their movements are generally very slow or more typically they are almost stationary for several weeks. These semi-permanent high pressure regions will now be shown to assume great importance for they are the breeding grounds of air masses.

CHAPTER 2

Air Masses — the Polar Front

THE central areas of the semi-permanent sub-tropical and polar high pressure regions are areas of 'quiet' weather, i.e. light winds and little or no rain. As a result of the light winds air remains within these pressure features for weeks or even months at a time before it leaves their periphery as described in Chapter 1. During this stagnant period air close to the earth's surface, everywhere within the high pressure regions, acquires the general temperature and humidity which are characteristic of these regions, i.e. air within the sub-tropical highs is warm while that within the polar high is cold; in both high pressure belts, if the source region lies over land the air will be dry, and conversely, over the sea the air will be humid. The vast quantities of air within the semi-permanent highs, having almost uniform temperature and humidity appropriate to their source region, are known as Air Masses.

On leaving its source region the air becomes modified by the surface over which it travels. Air moving southwards is warmed, while air moving from the continents out over the oceans becomes more humid and vice versa.

Air masses are classified by their source regions and whether these sources lie over land or water. The main masses are thus labelled:

Tropical Maritime Air
Tropical Continental Air
Polar Maritime Air
Polar Continental Air.

The most frequent locations of these main air masses are shown in Diagrams 8 and 9. These source regions are the semi-permanent high pressure areas shown in Diagrams 4 and 5. Since the latter differ from the idealized pattern of Diagram 2, we shall now have to look more closely into the location of the semi-permanent highs.

The variability in the position of the polar and sub-tropical high pressure areas has already been briefly referred to in Chapter 1. The locations of these highs are affected by the distribution of continent and ocean and by the differing seasonal temperatures over the continents. Diagram 2 is a schematic representation of the average annual pressure pattern over a uniform (no land/sea differences) globe. A more realistic picture is given in Diagrams 4 and 5, representing the summer and winter condition over the greater part of the globe. (Spring and autumn can be considered transitional periods between the other seasons.) It should be understood that these diagrams show the average position of the semi-permanent highs and lows; the features themselves may be found

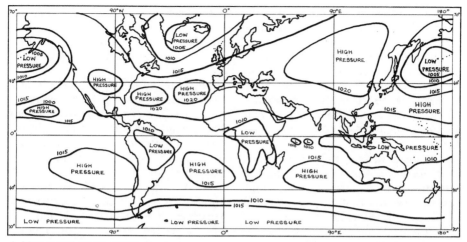

4. Mean monthly pressure (mb) – January

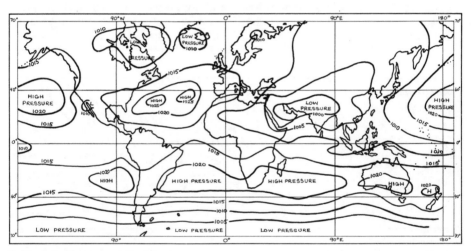

5. Mean monthly pressure (mb) – July

on particular occasions as far as one or two thousand miles from their average positions.

The equatorial low pressure belt and the sub-tropical highs in each hemisphere can be recognized in both seasons. The northeast and southeast trades, originating from the sub-tropical highs, converge towards the equatorial region. The trade wind air masses (tropical maritime air) are typified by blue skies with scattered fleecy rainless clouds (trade wind cumulus), but the weather can some-

times be disturbed for the trade wind belt is the breeding and subsequent hunting ground of tropical revolving storms. Within the equatorial region the winds become much more variable and the weather often thundery - the typical condition of the Doldrums. (This region is technically referred to as the Inter-Tropical Convergence Zone (see Plate section); it moves a few degrees north and southwards following the apparent movement of the sun.) The air mass over the Doldrums, originally trade wind air

19

but considerably modified by its journey over a warm ocean, is often referred to as Equatorial Air.

The sub-tropical high pressure belts are not continuous around the globe. Diagrams 4 and 5 show that the highs are chiefly confined to the oceans. They also show that considerable seasonal changes occur over the continents.

The most striking difference occurs over Asia. In winter, due to the low temperatures of the region, high pressure becomes established over this continent - the Siberian High (a part of and often the main polar high of the northern hemisphere). In summer when Asia becomes warm, the Siberian High is replaced by a large area of low pressure centred over northern India. This seasonal change of pressure has a considerable effect on the wind pattern over the whole of southeast Asia including the Arabian and China Seas and the Bay of Bengal. Here, in winter, the winds are northeasterly (part of the clockwise flow around the Siberian High); in summer the winds are southwesterly (part of the counter-clockwise flow around the Asian Low). These winds are known as the North-east and Southwest Monsoons (a monsoon being seasonally reversing winds). Less dramatic monsoons also occur in some other parts of the tropics, notably West Africa and northwest Australia (see Diagrams 6 and 7). The winds of the monsoon regions are normally blowing from ocean to continent so that the air masses are hot and humid (equatorial air), resulting in heavy rainfall. A main exception is the northeast Monsoon which blows from continent to ocean and is usually very

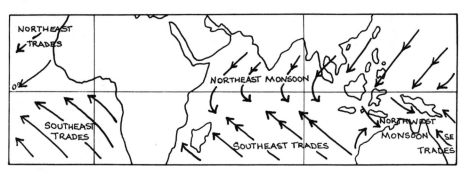

6. Generalized winds over the tropical oceans (40°W to 150°E) – January

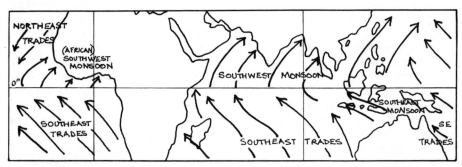

7. Generalized winds over the tropical oceans (40°W to 150°E) – July

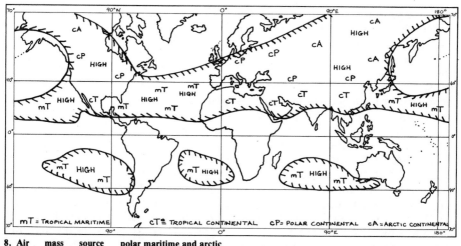

8. Air mass source regions – January. Due to the limits of latitude (70°) the main sources of polar maritime and arctic maritime air cannot be shown.

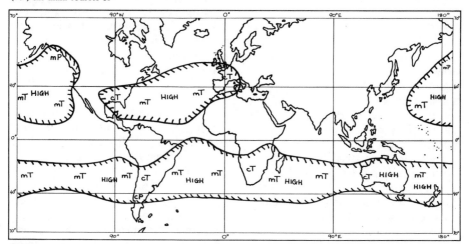

9. Air mass source regions – July. (For legend and note, see Diagram 8.)

dry (tropical continental air) over most of this region, apart from Malaysia and Indonesia where there is copious rainfall due to the maritime track of the Northeast Monsoon upwind from these areas (see Diagram 6). We shall now leave the air masses of the trade wind and monsoon regions and return to those of the region between the sub-tropical and polar highs in each hemisphere.

Diagrams 8 and 9 show the chief source regions of the four main air masses: tropical maritime (mT), tropical continental (cT), polar maritime (mP) and polar continental (cP). In the northern winter when the Arctic is very cold and the sea-ice extends much farther south than in summer (Diagram 79) the bitterly cold air moving south from this region, sometimes as far as industrial Northwest Europe, is known as an Arctic air mass. The Antarctic air mass is usually

21

confined to that continent but may occasionally break northwards reaching the tip of South America.

The air mass at a particular place will depend on the relative position of the source region and also on the positions of travelling depressions. A depression approaching western Europe from the southwest will be preceded by southeast to southwest winds; these will normally have a fetch from the sub-tropical high pressure area in the vicinity of the Azores - these air flows are tropical maritime air. As the depression passes by to the north and the wind becomes northwesterly, the air being drawn around the rear of the depression will have originated in polar latitudes and since it will have travelled over the ocean it will be polar maritime air.

In winter, should a further veer produce a strong northerly, air will be drawn directly from the pack-ice regions of the Arctic - an Arctic air mass. This air mass is not frequent south of about 60°N and very rare south of about 50°N over western Europe.

High pressure over Scandinavia in winter usually results in a northeasterly or easterly airstream over the UK and adjacent western Europe introducing polar continental air, and so on.

As an example for a southern hemisphere location, the main air masses affecting the Falkland Islands (approx 51°S, 60°W) are polar maritime and tropical maritime air due to the constant procession of eastward-moving depressions passing close to this region. Occasionally, and more often in the southern summer (January), a persistent northerly wind may bring tropical continental air from South America. In deep winter (July) a prolonged southerly will bring very cold Antarctic air into the region.

Over the Caribbean, where the northeast trades predominate, the main air mass is tropical maritime air

(trade wind air). In winter a cold northerly sometimes develops, especially over the western Caribbean; this northerly brings polar continental air.

The Baltic, in summer, will most often experience tropical continental air and less frequently polar and tropical maritime air depending on depression tracks.

The Mediterranean in summer is itself an air mass source region (tropical maritime air). Pressure is relatively high and winds, over the open sea, are usually light. However, near the mainland coasts persistent offshore winds will bring continental, dry air over the sea. In winter polar maritime air is often drawn into the Mediterranean in the rear of eastward-moving depressions over southern Europe. The resulting situation, cold air over a relatively warm sea, favours the formation of depressions, preferentially in the Gulfs of Marseilles and Genoa, often producing violent weather.

The main weather characteristics of air masses will be dealt with more fully in Chapter 8. It will suffice to note here that continental air whether polar or tropical is normally dry and produces little or no 'weather' whereas maritime air is normally humid and produces cloud and some rain. Further, polar maritime air produces intermittent showers of rain (or snow depending on season), often alternating with sunny periods, whereas tropical maritime air normally produces generally damp, cloudy weather, sea and hill-fog and drizzle, more especially in winter, spring and early summer.

We shall see later on (Chapter 8) that most 'weather' (e.g. continuous and often heavy rain or snow) occurs at the boundaries between air masses and not within an individual air mass.

Diagram 3 shows the Polar Easterlies and the Disturbed Westerlies (more correctly Southwesterlies) converging into the latitude belt between roughly 50° and 60° in each hemisphere. These air masses, the one of polar origin and the other of tropical origin, do not mix at their common boundary, due to their differing temperatures (densities). Instead the tropical air, warmer and less dense, rises gently over the colder polar air forming a zone of cloud and weather (rain, etc). This zone is known as the Polar Front, the battleground between tropical and polar air, each struggling to extend its respective journey northward or southward.

The Polar Front can be imagined to extend around the globe in the temperate latitudes, at least in idealized form. In fact, over the continents where the air is normally dry and where there are considerable differences in the position of the sources of polar and tropical air, the boundary is often difficult to trace; and for the purposes of this book, anyhow, is unimportant. The Polar Front is normally fairly easily detectable over the oceans. A mean position is not shown diagrammatically as it is normally constantly moving within wide limits, for reasons which will be discussed in the next chapter. In general it can be said that it lies further south, on average, in the western North Atlantic (at about 30°N on the American seaboard) and at about 50°N in the east. The position is similar in the North Pacific; here the Polar Front, on average, lies from Taiwan to the Aleutians.

The Polar Front in the southern hemisphere, though more easily detected than in the north, also moves over a wide latitude belt ranging from about 30° to around 60°S. Its mean position lies between 50° and 60°S.

In both hemispheres the Polar Front moves a few degrees of latitude north and south with the changing seasons, and becomes less active in summer when the temperature difference between the tropical and polar air masses is decreased.

The region of the Polar Front in both hemispheres is meteorologically very important since it is the chief breeding ground of the travelling low pressure areas called depressions.

CHAPTER 3

Formation of Depressions on the Polar Front

THE average positions of the 'semi-permanent' low pressure regions in winter and summer are shown in Diagrams 4 and 5 in the previous chapter. Over the southern hemisphere, more or less uncomplicated by the presence of land in the relevant latitudes, the low pressure area is shown as a belt around the globe in about 50° - 70°S. This belt represents a stream of depressions, separated perhaps by a thousand miles or so, all moving generally eastwards just off the coast of Antarctica. (The strong westerly wind belt between these lows and the sub-tropical high pressure belt at about 30°S is commonly referred to as the 'Roaring Forties', the 'Furious Fifties' or the 'Shrieking Sixties' according to latitude.)

There is no such semi-permanent low pressure belt in the northern hemisphere. The low pressure areas shown over the oceans reflect the mean position of depressions which, having formed on the western sides of the oceans, develop into full maturity in about the positions shown in Diagrams 4 and 5, and subsequently decay over the continents on the eastern side of the oceans. The average positions give rise to the labels Icelandic and Aleutian Lows; individual lows are just as frequently found in other longitudes in this belt, but then they are usually not as well developed.

In both hemispheres, then, it can be seen that depressions, at least over the oceans, are generally mobile: hence the term 'travelling depression'. Depressions are also mobile over land though their speed is usually reduced by the roughness of the land surface. The major low pressure areas forming over the continents through heating in summer are almost stationary, while the equatorial low pressure area (the Doldrums) is simply a belt of lower pressure, without definite circulation, between the sub-tropical highs of each hemisphere.

Depressions are depicted on weather charts as series of more or less concentric isobars in the same way that contours define hills and mountains on topographical maps. (An isobar is a line joining places of equal atmospheric pressure.) Pressure is most commonly measured in millibars (see Appendix 1), and isobars are normally labelled with these units. A typical central value for a 'deep' depression would be 960 millibars (mb) and for a large anticyclone (high pressure area) about 1040 mb; extreme values for surface pressure would be about 30 mb outside this range.

Most depressions form on the Polar Front in each hemisphere. Stages in their development are shown

10. The formation of a depression on the Polar Front – northern hemisphere. Stage 3 is perhaps 12–24 hours after Stage 1.

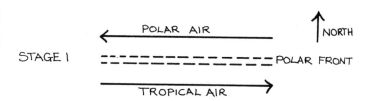

STAGE 1

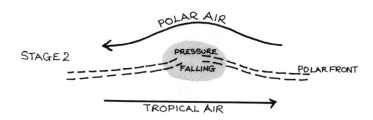

STAGE 2

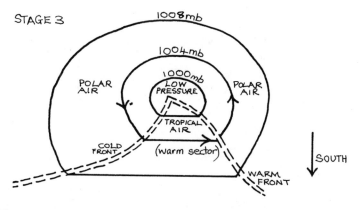

STAGE 3

schematically in Diagrams 10 and 11. Stage 1 of Diagram 10 (northern hemisphere) shows tropical air flowing eastwards alongside westward-moving polar air with the Polar Front at their common boundary. The process of developing a depression on the Polar Front is aided by a strong temperature contrast between the polar and tropical air masses. These conditions are most commonly met on the western sides of the oceans in winter.

Perhaps, due in part to friction between the two air masses, a wave-type distortion appears on the Polar Front (Stage 2) with a local fall of pressure at the tip of the wave. On many occasions such waves show no further development but simply run eastwards (under the influence of westerly winds in the upper air) along the Polar Front to decay in a day or two.

However some waves develop into vigorous depressions; the 'deepening' process is largely due to horizontal

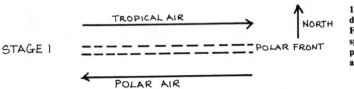

STAGE 1

TROPICAL AIR →

NORTH ↑

- - - - - - - - - - - - - POLAR FRONT

← POLAR AIR

11. The formation of a depression on the Polar Front – southern hemisphere. Stage 3 is perhaps 12–24 hours after Stage 1

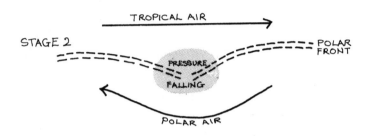

STAGE 2

TROPICAL AIR →

POLAR FRONT

PRESSURE FALLING

←

POLAR AIR

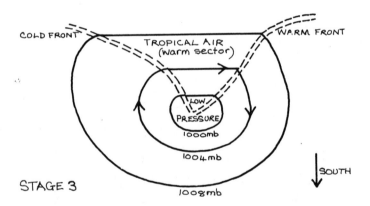

COLD FRONT

TROPICAL AIR (Warm sector)

WARM FRONT

LOW PRESSURE

1000mb

1004mb

1008mb

SOUTH ↓

STAGE 3

temperature differences at heights between about 5 and 10 km. Stage 3 shows such a depression with several isobars drawn in around its centre: the air rotating in a counter-clockwise sense. It will be noted that the Polar Front is now markedly distorted.

This is a typical depression with its own 'captured' portion of the Polar Front which will move along with the depression. The tropical air lies between the two portions of the front (Stage 3). That portion on the right of the tropical air is known as a Warm Front and that on the left as a Cold Front. (Fronts are more fully discussed in Chapter 8.) The tropical air between the fronts is known as the Warm Sector of the depression. For the greater part of the life of a depression the warm sector will remain on its equatorward side.

Depressions in the southern hemisphere form on its Polar Front in exactly the same manner. Due to the fact that the eastward-moving tropical air lies on its northward (equatorward) side, the formation of a depression in

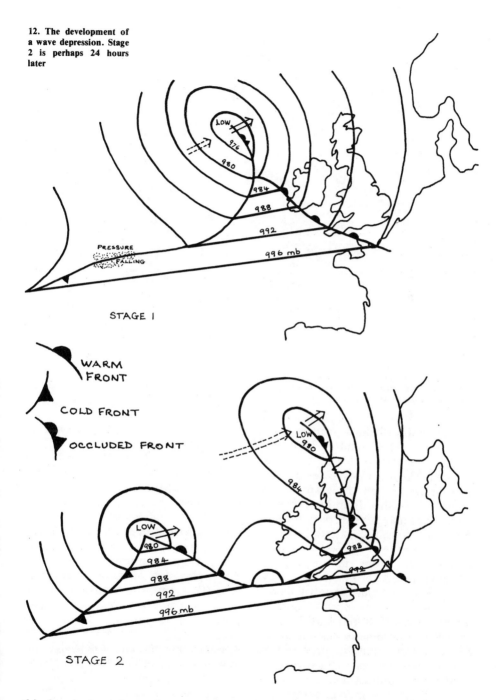

12. The development of a wave depression. Stage 2 is perhaps 24 hours later

STAGE 1

WARM FRONT

COLD FRONT

OCCLUDED FRONT

STAGE 2

this hemisphere is an inverted image of that in the northern hemisphere (Diagram 10); the resultant circulation is clockwise.

In both hemispheres the development process may be repeated on the Polar Front at about 1500 km or so to the west of the original depression which will have moved in a generally easterly direction and driven the Polar

27

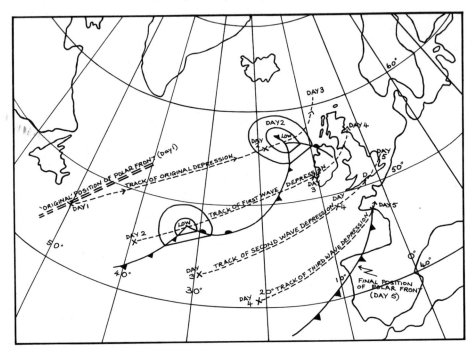

13. A 'train' of wave depressions. (The illustrated depressions are shown for a time soon after the beginning of Day 2.)

Front (in this case the depression's own cold front) a few hundred kilometres on the equatorward side of its normal position. The new low is categorized as a Wave Depression (Diagram 12 and Plates). A further four or five wave depressions may subsequently form on the Polar Front, the 'easterly' track of each lying a few degrees of latitude equatorward from its predecessor (Diagram 13). After this train of wave depressions the Polar Front will lie well equatorward of its normal position (perhaps over 1500 km) and the polar air will have become so much modified by warming from a progressively warmer land or sea surface that the temperature contrast across the Polar Front will have become ill-defined and the Polar Front will decay. Within a few days thereafter the tropical and polar air masses normally re-group and the Polar Front becomes re-established close to its mean position so the scene is then set for the process of depression formation to begin anew.

It will be seen from Diagram 12 that a different type of front has been drawn in where the warm and cold fronts meet; this is called an Occluded Front or, more commonly, an Occlusion (see Chapter 8). A third type of depression, a Secondary (or Break-Away) Depression sometimes forms at the 'triple-point' where the three fronts meet (Diagram 14). (Though this is not by any means a rare event, most triple-points fail to produce a secondary depression.) Pressure sometimes falls at the triple-point and a new vigorous low may result which may completely swamp the parent low. This process may be sudden so that the transition from the two stages shown in Diagram 14 may take place in 12 to 24 hours. Secondary depressions, together with their acquired fronts, normally move east.

14. The development of a secondary (break-away) depression. Stage 2 is perhaps only 12 hours later.

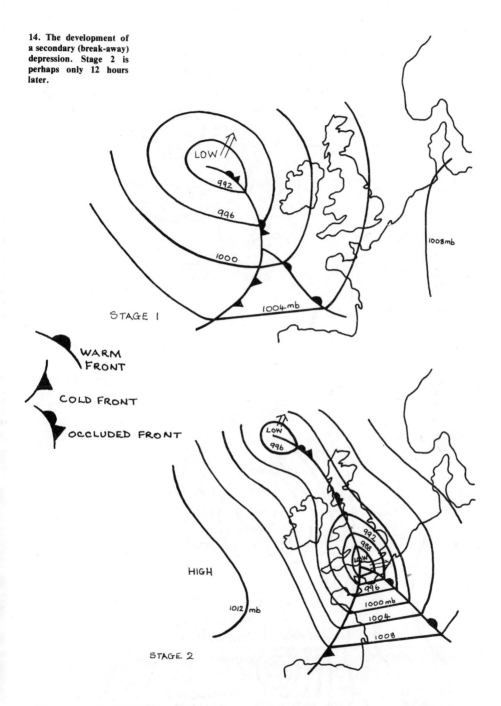

STAGE 1

WARM FRONT

COLD FRONT

OCCLUDED FRONT

STAGE 2

Diagrams 4 and 5 indicate that large-scale low pressure areas form over the continents from surface heating in summer. Similar smaller depressions may form almost anywhere overland after a few days of prolonged summer heating; they are called Heat Lows. The wind circulation around such lows is usually light and their main significance lies in the fact that they are

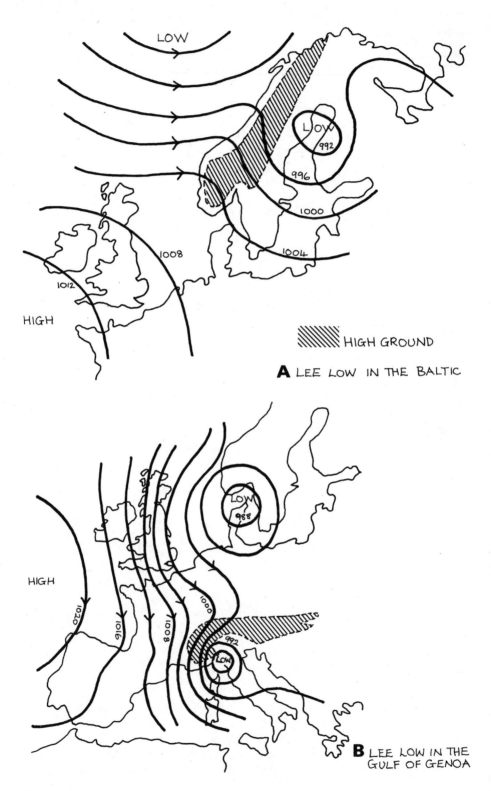

LOW

LOW
992

996

1000

1004

1008

1012

HIGH

HIGH GROUND

A LEE LOW IN THE BALTIC

HIGH

1020

1016

1008

1000

LOW
988

992

LOW

B LEE LOW IN THE
GULF OF GENOA

usually areas of thundery weather. Over western Europe, heat lows are commonly found over Spain and France and less frequently over the UK.

Another type of depression is a Lee Low, which as its name suggests forms in the lee of a large mountain barrier lying perpendicular to a fairly strong wind. It is a fact that air flowing over a mountain range will acquire a 'cyclonic twist' on descending over the lee side. This may be sufficiently developed to result in a complete depression in the lee of the mountain range. (A cyclone is the generic name for any low pressure area.) In the European sector such lows readily form in the Gulf of Genoa in the lee of the Alps, and a strong westerly wind over Scandinavia may generate a lee low over the Baltic (see Diagram 15).

A final type of depression, called a Polar Low, is described in Chapter 10. Tropical storms (hurricanes, typhoons etc) will be discussed in Chapter 17.

15. The formation of lee lows (Scandinavia and Gulf of Genoa)

CHAPTER 4

Wind Directions Around Depressions and Anticyclones

THE direction of wind flow around depressions is anticlockwise in the northern hemisphere and clockwise in the southern; around an anticyclone it is clockwise in the northern hemisphere and anticlockwise in the southern. These directions are invariable.

We shall now discover why this should be so, and also why the air flows around these features and not directly into a depression or out of an anticyclone.

First we shall assume that within a given test area the pressure is everywhere the same, i.e. there are no isobars and therefore there is no pressure gradient. The individual air particles are at rest as there are no forces acting on them. (This is strictly true only in the horizontal direction; in the vertical, gravity is balanced by the buoyancy of the air: but we are here concerned only with horizontal forces and consequent horizontal motion.)

A pressure gradient is now applied in our selected area, as would happen at the initial stage of development of a depression, indicated in Diagram 16 by the presence of some isobars such that, for the purposes of this example, lower pressure is located towards the top of the page and higher pressure towards the bottom. As a consequence of introducing a pressure gradient we

have also introduced a Pressure Gradient Force which tends to move the air particles directly from high pressure to low pressure in order to neutralize the pressure gradient. The Pressure Gradient Force acts at right angles to the isobars, from high to low pressure, and is inversely proportional to the distance between the isobars, i.e. the closer the isobars (the steeper the pressure gradient) the stronger the Pressure Gradient Force. The air particles begin to accelerate directly towards lower pressure under the action of this single force.

But these particles are moving over a rotating earth and so as soon as they begin to move they are influenced by this rotation, or rather by a force which is due to the earth's rotation the Geostrophic Force, sometimes called the Coriolis Force. The effect of this force may be demonstrated on a schoolroom black globe, or on a tough rubber ball impaled through its 'north and south poles' on a knitting needle. Draw several small circles at different latitudes and mark the centre and any point on the circumference of each circle. Looking down on the north pole, slowly spin the globe anticlockwise, identically to the earth's own rotation. You will notice that in the case of circles near the pole the point on the circle spins around its

16. The geostrophic wind – northern hemisphere

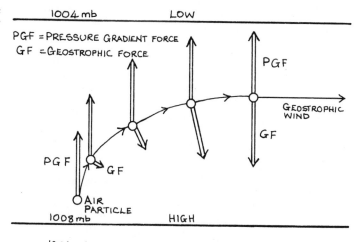

17. The geostrophic wind – southern hemisphere

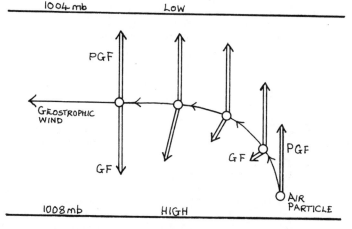

centre but the nearer a circle is to the equator the less is this 'local spin' observed. The direction of the spin is anticlockwise. Now invert the globe to look down on the south pole and rotate the globe in the correct direction - towards the east - now seen to be clockwise. You will again observe maximum local spin near the south pole and minimum towards the equator, but the local spin is now clockwise.

Any disc on the earth's surface has local spin, maximum at the poles and zero at the equator. We now consider a balloon-sized 'parcel' of air setting off from the centre of any disc (other than at the equator) in the direction of the marked point on the circumference. The parcel is not 'fixed' to the earth.

When it has moved in a straight line only a little distance, the circumference point has already moved away (in the northern hemisphere) anticlockwise, to the left, and *to an observer standing on the rotating earth* the balloon of air will have deviated to the right. In the southern hemisphere the parcel will have deviated to the left.

Thus the Geostrophic Force acts to the *right* of the air's motion in the northern hemisphere and to the *left* in the southern. It has been shown that it is proportional to latitude, being a maximum at the poles and zero at the equator. Further, it is proportional to the speed of the moving air, i.e. as the air's speed increases so the Geostrophic Force increases.

To return to Diagram 16 (the northern hemisphere case), an air particle accelerating towards lower pressure under the Pressure Gradient Force is progressively deflected to the *right* by the steadily increasing Geostrophic Force. While the Pressure Gradient Force remains the greater force the air particle will continue to accelerate but as its speed increases so its deflection also increases. Consequently the direction of the Geostrophic Force is steadily becoming more directly opposite to that of the Pressure Gradient Force. A balanced state is finally reached when the Geostrophic Force acts so as to directly oppose the Pressure Gradient Force and is equal in magnitude to it. Since the latter always acts at right angles to the isobars this can only be attained when the resulting air movement (the wind) is parallel to the isobars, as

18. Wind directions around depressions

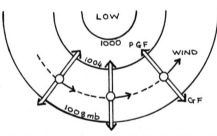

18a. Northern hemisphere. PGF always acts directly towards centre of low pressure. GF acts to	*right* of air movement. Resultant wind blows anticlockwise around low.

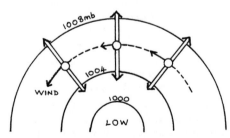

18b. Southern hemisphere. PGF always acts directly towards centre of low pressure. GF acts to	*left* of air movement. Resultant wind blows clockwise around low.

19. Wind directions around anticyclones

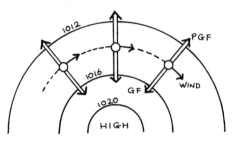

19a. Northern hemisphere. PGF always acts away from centre of high pressure. GF acts to *right*	of air movement. Resultant wind blows clockwise around high.

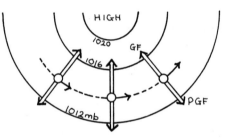

19b. Southern hemisphere. PGF always acts away from centre of high pressure. GF acts to *left*	of air movement. Resultant wind blows anticlockwise around high.

shown in the diagram. Though there is then no net horizontal force acting on the air, the air will continue to move because it is already moving. (This accords with Newton's First Law of Motion.) In the example of Diagram 16 (northern hemisphere) the air continues to move parallel to the isobars so that the lower pressure lies on the left.

In the southern hemisphere, the Geostrophic Force acts towards the *left*. The progression towards balanced flow is shown in Diagram 17. Here the low pressure lies on the right.

In each case, the derived 'steady-state' wind is called the Geostrophic Wind.

These two cases give us Buys Ballot's Law (practically the only invariable law in the whole field of meteorology), that if one stands with one's back to the wind then the low pressure lies on the *left* hand in the northern hemisphere and on the *right* in the southern.

The foregoing has been related to straight isobars. When the isobars are curved the same results will be found. In Diagram 18 the isobars are concave towards the low pressure, as they are in a depression: the resultant wind will flow parallel to the isobars in an anticlockwise direction in the northern hemisphere and in a clockwise direction in the southern hemisphere.

With isobars concave towards high pressure, as in an anticyclone, the wind flow around a high will be clockwise in the northern and anticlockwise in the southern hemisphere (Diagram 19).

Another force is involved when dealing with curved isobars. This is the Cyclostrophic Force which acts towards the centre of curvature and is proportional not only to the speed of the air but also the radius of curvature. The Cyclostrophic Force has no effect on the direction of the wind but it does reduce the wind speed around a depression and increase it around an anticyclone, in both hemispheres. In a typical depression this decrease will amount to about 20 per cent of the wind speed for the straight isobar case: though in an extreme case such as a hurricane or typhoon where the pressure gradient is particularly steep and the radius of curvature correspondingly small, it can amount to a 75

20. The surface wind – northern hemisphere

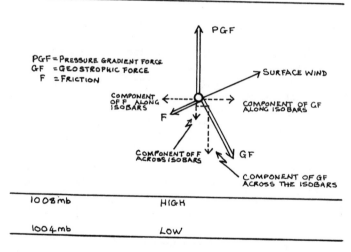

21. The surface wind – southern hemisphere

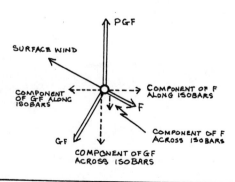

per cent reduction. (Even so winds in hurricanes and typhoons still sometimes approach 200 knots!) Typical increases for an average anticyclone are much smaller amounting to about 10 per cent. In contrast to the term Geostrophic Wind which is the wind related to straight isobars, the wind associated with curved isobars is technically referred to as the Gradient Wind.

In the absence of friction between the moving air and the earth's surface, the Geostrophic (or Gradient) Wind would be the wind actually experienced at the surface - the surface wind. However the effects of friction cannot be ignored and since they extend to a height of about 900 m above the earth's surface, the Geostrophic and Gradient Winds are representative of the wind at that level even though they are related to isobars of *sea-level* pressure.

So in order to derive the surface wind, we must now return to Diagrams 16 and 17 and introduce yet another force, that due to friction. This force acts in a direction opposite to the air movement and is proportional to the speed of the air. As three forces are now involved, balanced flow is reached when the components of each force acting firstly parallel to the isobars and next perpendicular to them cancel each other in each direction. This balanced state is reached before the steady state shown in Diagrams 16 and 17. The forces acting and their components are shown in Diagrams 20 (northern hemisphere) and 21 (southern hemisphere). In Diagram 20 the total sum of the components of the Geostrophic Force and Friction acting across the isobars equally oppose the Pressure Gradient Force. In the direction parallel to the isobars, the component due to the Geostrophic Force is exactly balanced by that of friction (the Pressure Gradient Force cannot have a component perpendicular to itself).

All the forces are in balance (since their perpendicular components are in balance) and the resultant wind, now the true surface wind, blows slightly across the isobars from high to low pressure. It is in fact backed (an anticlockwise change) by about 30° from the direction of the isobars. Diagram 21 presents the case for the southern hemisphere. Here balanced flow occurs with the surface wind veered (a clockwise change) by about 30° from the direction of the isobars.

Applying these arguments to depressions and anticyclones, the resultant surface wind patterns are as shown in Diagrams 22 and 23. In both hemispheres it can be seen that at the surface there is a net inward flow of air into a depression and a net outward flow from an anticyclone. This inflow and outflow alone would very soon (i.e. within a matter of hours) fill a depression or 'deflate' an anticyclone by destroying the original pressure gradients. They do not do so because both processes, inflow and outflow,

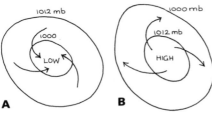

22. The surface wind pattern around depressions (a) and anticyclones (b) in the northern hemisphere

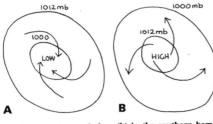

23. The surface wind pattern around depressions (a) and anticyclones (b) in the southern hemisphere

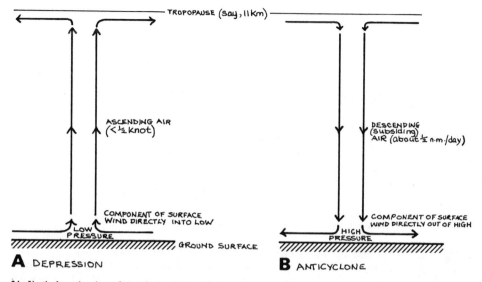

24. Vertical motion in a depression and an anticyclone

are compensated by vertical motion within depressions and anticyclones.

The air inflowing at the surface slowly rises through the central regions of a depression (at usually much less than ½ knot) and finally flows out at the top, say at about 11 km high (Diagram 24a). The outflowing air at the base of an anticyclone is the result of descending air within the high (at a rate even slower than ascent in a low, say ½ NM/day) which in turn results from inflow at the top of the high (Diagram 24b).

It is thought that the outflow and inflow at the top of depressions and anticyclones, respectively, in opposition to the inflow and outflow at their bases, are caused by horizontal temperature differences at the upper level and that these temperature differences are a fundamental cause of the development of depressions and anticyclones.

While the outflow at the top of a depression exceeds the inflow at its base the depression is deepening (that is surface pressure is falling within the depression), and when inflow exceeds outflow the depression fills. The

deepening stage of a depression is accompanied by worsening weather (heavier and more continuous rain as well as stronger winds) due to increased vertical motion and a resulting steeper pressure gradient; while at the filling stage, due to decreased vertical motion and the resulting weaker pressure gradient, the rain decreases in intensity or may even die out and the winds are less strong.

Conversely, with an anticyclone as long as inflow at the top exceeds outflow at the bottom, the anticyclone will intensify (or build). When inflow exceeds outflow the anticyclone weakens.

Rising air leads to condensation, clouds and rain etc, therefore depressions are commonly regions of bad weather. Descending air leads to evaporation of clouds resulting in fine weather; anticyclones are typically areas of fine weather.

Just how this comes about will be discussed in the following few chapters which describe clouds and their formation and the buoyancy (stability) of the atmosphere.

CHAPTER 5

Moisture and Stability in the Atmosphere

MOISTURE in the atmosphere is normally present as water vapour, an invisible gas which constitutes as an absolute maximum only 4 per cent of the mixture of gases known as air. Yet this small constituent provides the earth's cloud cover and its rainfall.

Most of the rain returns to the oceans, lakes or ground, from whence evaporation occurs to replace the moisture in the air. This is commonly referred to as the Water Vapour Cycle.

Water vapour is always present to a greater or lesser extent at least over the lowest layers of the troposphere. There is a maximum amount of water vapour which can be mixed into a given volume of dry air at a given temperature; when reached the air is said to be Saturated. The water vapour is still invisible in saturated air. In fact, without instruments to measure humidity, it would be difficult to realize that the air was saturated.

The further addition of water vapour into an already saturated volume of air, say by evaporation from rain falling through it, would immediately result in the excess being condensed out as very small, but very visible, water droplets which collectively form cloud.

The amount of water vapour required to saturate a given volume of dry air depends only on the temperature of the air. As examples, saturation of a given volume of dry air at 10°C (50°F) occurs when water vapour occupies 1 to 1½ per cent of the volume occupied by the dry air; at 25°C (77°F) the corresponding value is 3 per cent.

To assess the degree of saturation, or the humidity, of the air, its water vapour content has to be related to its temperature. In practice the humidity is frequently expressed as a related percentage ratio in the following form:

$$\text{Relative Humidity} = \frac{\text{The amount of water vapour present}}{\text{The amount required to saturate the air at the same temperature}} \times 100$$

Obviously 100 per cent relative humidity means saturation. As an indication of the normal range of values, at least in temperate latitudes, air with a relative humidity below 60 per cent is classified as 'dry air'. Values below 20 per cent are extremely rare at sea level in the UK.

Another measure of the humidity of the air is its Dew-point Temperature, defined as the temperature of the air at which, on cooling, condensation (dew)

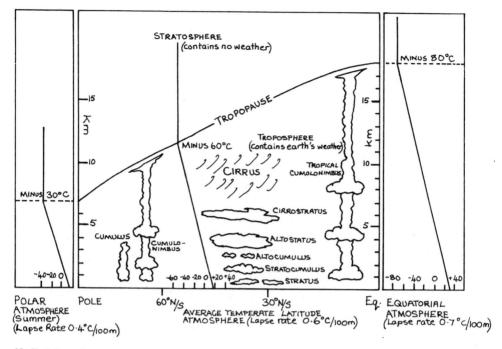

25. Variation of tropopause height with latitude: associated lapse rates

first begins to form. The dew-point temperature is also a related term; it must be related to the actual temperature of the air (the dry-bulb temperature). Thus if we know that the air temperature and dew-point temperature are both, say; 12°C (54°F) we know that the air is saturated; if the air temperature is 12°C (54°F) and the dew-point temperature is 5°C (41°F) then the air is in an unsaturated state and relatively dry.

Water vapour decreases with increasing height and is almost entirely confined to the troposphere. Clouds and weather are therefore confined to this relatively shallow layer of the atmosphere, within which temperature decreases with increasing height. As far as the troposphere is concerned the heat source is the earth's surface, which is heated by the sun. The highest temperatures therefore occur near to the earth's surface and the lowest temperatures at the top of

the troposphere - the tropopause (Diagram 25). The rate of decrease, or lapse of temperature (the Lapse Rate), is constantly varying. The average lapse rate is about 0.6°C/100m. At this value the buoyancy of the air (a measure of its resistance to vertical motion) is about neutral; i.e. air displaced upwards or downwards, for any reason, will tend to stay in its new position.

Should this neutrally buoyant air be moved horizontally over a warmer ground surface it will be warmed from below and the lapse rate will increase. Rising air, then, would probably continue to be warmer (less dense) than the environment air at each level passed through; its ascent would be maintained until its temperature was equal to that of the environment when it would lose its buoyancy and cease to rise (Diagram 26). In nature, air heated from below rises in discrete 'bubbles', perhaps a few hundred feet across, just

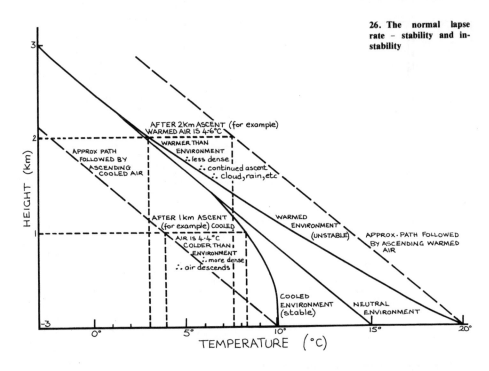

26. The normal lapse rate – stability and instability

HEIGHT (km) / **TEMPERATURE (°C)**

AFTER 2km ASCENT (for example) WARMED AIR IS 4·6°C

WARMER THAN ENVIRONMENT ∴ less dense ∴ continued ascent ∴ cloud, rain, etc.

APPROX PATH FOLLOWED BY ASCENDING COOLED AIR

AFTER 1km ASCENT (for example) COOLED AIR IS 4·4°C COLDER THAN ENVIRONMENT ∴ more dense ∴ air descends

WARMED ENVIRONMENT (UNSTABLE)

APPROX. PATH FOLLOWED BY ASCENDING WARMED AIR

COOLED ENVIRONMENT (stable)

NEUTRAL ENVIRONMENT

as bubbles rise in boiling water. This process is called Convection and the atmosphere is said to be Unstable, i.e. it is conducive to upward motion.

If we return to our sample of neutrally buoyant air and move it over a cooler surface, it will be cooled from below and its lapse rate would then become less than the normal 0.6°C/100m. Our sample of air, on being forced to rise, would be colder (more dense) than the environment air at the upper level and would fall to its original level (Diagram 26). The atmosphere is now Stable, i.e. it resists upward motion.

In general, then, a polar air mass moving equatorward (warmed from below) becomes unstable. Conversely, tropical air moving poleward (being cooled from below) becomes more stable.

This is a much simplified account of the temperature structure in the troposphere and its effect on vertical motion. Nevertheless, it indicates the processes involved. In practice, the lapse rate is not uniform throughout the depth of the troposphere at any given time. It also varies with time, often in a few hours, due to the complicated behaviour of the troposphere.

Though the temperature normally decreases with height, there are occasions when the reverse is true and the temperature increases with increasing height. This reversal of the normal lapse of temperature is called an Inversion (see Diagram 27).

Inversions at the bottom of the troposphere readily form overnight whenever there are clear skies and light winds - the ideal conditions for maximum cooling of the ground surface. They also occur at fronts, where warm air overrides cold air along a sloping surface, as we shall see in Chapter 8. The most marked inversions are commonly those which are due to descending air, typically within an anticyclone. The process of descent is known as Subsidence; it begins at the upper levels of an anticyclone.

27. An inversion of temperature

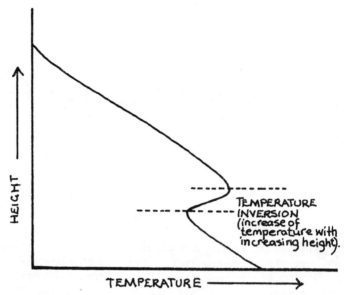

28. The formation of an inversion by subsidence

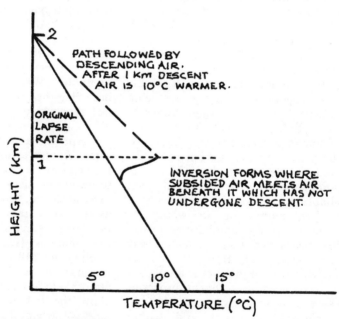

Subsiding air, falling through levels of increasing pressure, is automatically compressed and warms up (the process is analogous to the familiar warming of a bicycle pump). The rate of warming is greater than the typical lapse rate of ascent (it is about 1°C/100 metres) so that after only a thousand metres of descent the air is 10°C warmer than the air immediately below it and an inversion results (see Diagram 28).

Subsiding air and consequent inversions (features of anticyclones) are associated with fine weather, but what of air which for one reason or another is caused to rise within the troposphere?

41

CHAPTER 6

Cloud Formation and Types

RISING air, because it is entering regions of increasingly lower pressure, expands and as a result it cools. (This cooling and the warming which occurs with descent are called adiabatic processes.) With continued ascent the air, originally unsaturated, will cool to its dew-point temperature and become saturated. Further cooling will lead to the excess water vapour now in the air being condensed out as tiny water droplets - cloud particles, a small fraction of a millimetre in diameter (see Table 1). The height to which air has to rise to attain condensation depends on its initial humidity. If this is already high (almost saturated air) then only a hundred metres or so would be sufficient; if the humidity is lower, then one or two thousand metres may be required; with absolutely dry air condensation can not occur at all.

There are four different processes which cause air to rise:

  Ascent at fronts and within
    depressions (mass ascent)
  Convection (surface heating)
  Orographic ascent (over high
    ground)
  Turbulence.

At fronts the warm (tropical) air, being less dense, rises over the cold (polar) air along the length of front. As a result the warm air is cooled and

condensation results in the formation of deep layers of cloud covering a large area (see Chapter 8). The process of air rising within the central regions of a depression has already been described in Chapter 4.

Convection results from strong heating of the air near to a warm ground; it makes the atmosphere more unstable. Where this process is occurring the ground temperature will not everywhere be the same. Due to different types of vegetation cover and soil type and also due to topography, some ground areas will have higher temperatures than others (especially south-facing slopes) and it is from these areas that 'bubbles' of air 'break away' from the surface and rise, perhaps in succession separated by a few minutes (Diagram 29). These bubbles are called Thermals and are familiar to glider pilots. If the atmosphere is already unstable it will allow the bubbles to rise to great heights, perhaps the whole depth of the troposphere (Diagram 25), forming in the process tall columns of cloud. Between these tall columns of cloud the air is gently descending to replace that undergoing convection.

Orographic ascent occurs when an upland region forms a barrier across the air flow at low levels. An isolated hill or mountain simply deflects the

29. Formation of convective cloud. Stages 1 and 7 are typically each separated by a few minutes.

2 km

1 km

1 2 3 4 5 6 7

1,2,3,4 Thermal 'bubbles' form on and then break away from the surface, expanding as they rise (pressure decreases with height).

5,6,7 Air within 'bubbles' cooled to dewpoint, thus condensation into cloud – initially 'fair weather' cumulus. If air is unstable larger cumu-

lus clouds form, perhaps joined by further thermal bubbles (6,7).

If air is very unstable, large cumulus grow into cumulonimbus, this

whole process taking perhaps no more than half an hour.

flow around itself but a line of hills, or better still a chain of mountains, forces the air to rise over it. Again, more than a light wind is required to produce the effect. The air on rising, of course, cools and, if saturation is exceeded, cloud will form. If the air is already fairly unstable then orographic ascent will result in discrete columns of cloud as in the convective process (Diagram 30). If the air is stable then a layer of cloud forms over, or on, the high ground. Given suitable conditions, the upland barrier may set up a vertical oscillation in the air flow which may result in five or six stationary 'waves'. At the crest of each wave a band of cloud may well develop (see Diagram 31); these bands are called wave (or lenticular) cloud and are common near upland regions.

Turbulence occurs in the atmosphere to a greater or lesser degree, e.g. the variations of the surface wind are largely due to turbulence. The degree of turbulence increases with the strength of the wind, and commonly produces upward and downward motions in the atmosphere, usually over a layer extending up to say 300 to 600m in depth. Moreover, wind

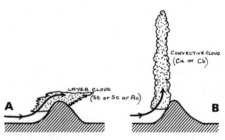

30. Orographic ascent cloud: stable (a) and unstable (b)

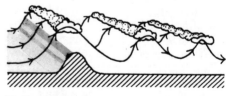

31. Orographic ascent – lee wave cloud

directions are not usually constant throughout the depth of the atmosphere; there are commonly large changes of direction sometimes in a shallow zone perhaps only a hundred metres in depth. The turbulence which results in these zones frequently produces a layer of cloud; if the air is already humid, the cloud will

43

commonly be of a patchy nature because the turbulence generates downward as well as upward motions, the latter producing the cloud and the former dispersing it.

Cloud Types

Clouds are classified as low, medium or high according to their composition, i.e. whether they are composed of water droplets, ice crystals or a mixture of both. This classification is really based on temperature. They are further classified as to whether they are layered (stratiform) or 'heaped-up' (cumuliform).

Whereas super-saturation is almost nonexistent in the atmosphere (once saturation is reached then any excess is almost immediately condensed out as water droplets), super-cooling of water droplets commonly occurs. Water droplets do not freeze as soon as their temperature falls below 0°C (32°F); in fact it is not until their temperature has dropped to below minus 20°C (minus 4°F) that most cloud droplets will have frozen. This temperature, on average, occurs at about 5,500m and this level is the lower limit for pure ice crystal cloud - high clouds.

Low clouds, composed of water droplets, are restricted to the layer of the atmosphere where the temperature is above 0°C (32°F); this temperature, on average, occurs at about 2,500m and this level is then the upper limit for low clouds.

Medium level clouds, which are composed of a mixture of water droplets and ice crystals, are found in the layer between 2,500m and 5,500m (temperature range 0°C to minus 20°C).

Since some deep cloud layers extend upwards across the thresholds, the heights quoted refer to the base of the cloud. (Further, as a technical point, since the heights given are global averages for these threshold temperatures they will of course vary, perhaps by a thousand metres or so, from time to time and from place to place.)

As a general rule when observing and identifying clouds, those which appear white, when viewed towards the sun, are almost certainly high clouds (ice crystal cloud) and those which appear grey contain water droplets; they are either medium level or low clouds. Again, as a general point, the darker the greyness the deeper the cloud. (All cloud types if viewed from the side away from the sun will appear white, sometimes brilliantly white, because they are good reflectors of sunlight.)

There are three types of high cloud:
1. Cirrus (Ci). Wispy, feather-like 'lines' of white cloud often called 'mare's tails'. Invariably the first cloud to be seen on the approach of fronts and depressions, but they are not always associated with them. Because of their association with fronts and more especially with depressions, they are, quite correctly, often taken as an indication that surface winds may increase in a few hours or so.
2. Cirrostratus (Cs). A whitish, milky 'veil', covering whole or part of the sky; it often shows haloes and is the only cloud to do so. Cirrostratus is again an indication of approaching fronts and depressions, but not always. The sun's outline appears rather hazy through cirrostratus.
3. Cirrocumulus (Cc). Small numerous white 'lumps' of cloud - an infrequent cloud type and the true 'mackerel sky'. Not of itself associated with any particular weather changes.
Cirrostratus and cirrus are formed by mass ascent. Cirrocumulus and to a lesser extent sometimes cirrus, are formed by turbulence.

At medium levels there are two types of cloud:

1. Altostratus (As). A grey, featureless layer of cloud covering the whole or most of the sky. It is most commonly associated with fronts and is frequently deep enough to produce rain or snow. (The featureless appearance is due to the rain falling from the cloud. At times the rain may evaporate entirely into the drier air below the base of this cloud, which will be at least 2,500m ft above ground level, and may not reach the ground.) Altostratus is caused by mass ascent and almost obscures the sun.

2. Altocumulus (Ac). Greyish-white close-packed 'lumps' of cloud, larger than cirrocumulus, most frequently caused by turbulence. Often called a 'mackerel sky' when viewed in the absence of other (higher) clouds. Not associated with any particular weather changes.

There are five types of low cloud:

1. Nimbostratus (Ns). A grey, featureless layer of cloud, completely obscuring the sun, commonly associated with fronts and depressions and from which prolonged rain (or snow) often falls. Nimbostratus is produced by mass ascent. It extends several thousand metres upwards from its base; the layer usually extends horizontally for hundreds of miles often merging with cirrostratus, especially in active fronts and near the centre of depressions to form one thick layer extending over almost the whole depth of the troposphere.

2. Stratus (St). A thin, grey, very low layer (typically below 300m), often very ragged in appearance, sometimes extending over the whole sky. It is produced by turbulence or orographic ascent.

3. Stratocumulus (Sc). Sometimes a complete cover of cloud showing 'rippling' due to small undulations in the height of the cloud base; it is commonly formed by turbulence. St and Sc are not indicative of changes in the weather.

4. Cumulus (Cu) (Plate section). Heaped-up cloud formed by convection and sometimes by orographic ascent. When tall columns of cloud having a cauliflower appearance are present it is a clear indication that the atmosphere is unstable and that showers may be expected. It is in these conditions that the surface wind become gusty and more than usually variable in direction (see Chapter 12). The base of cumulus cloud, normally viewed against the sun, appears very dark; when the side of the cloud is viewed in strong sunlight it appears brilliantly white.

5. Cumulonimbus (Cb). A large version of a cumulus cloud, it sometimes extends as a 'tall' column from its base at about 500m right to the tropopause. This can be about 9,000m in high latitudes and over 18,000m at the equator, where the width of the tall column may be 10 miles or more (Diagram 25). Cumulonimbus is the most dramatic and destructive of all clouds. Produced by vigorous convection, the up and down currents within this cloud are a hazard to aviation and we shall see in Chapter 12 that the down-currents produce sudden squalls which are a hazard to yachts. It is the only cloud which produces thunderstorms and hailstones. The base appears very dark indeed and very threatening, often with a hint of purple in its very dark grey colour. Again when viewed from the side in sunlight it will appear brilliantly white and it is then that its characteristic 'anvil' top may be seen. Distant cumulonimbus clouds take on a pale pink tint when illuminated by direct sunlight.

CHAPTER 7

Precipitation

MOST precipitation falls from relatively deep cloud formed by mass and orographic ascent or by convection. The rising air easily supports the weight of cloud droplets, since their terminal velocities are so small (Table 1), and so they are carried upwards by the ascending air. This chapter briefly describes the transition from cloud droplets to precipitation falling to the ground as drizzle, rain, sleet, snow and hail. Other more rare forms such as snow grains, ice pellets and ice needles need not concern us here in this brief account.

There are two accepted theories accounting for precipitation; they differ in that one requires a freezing stage and the other does not.

The Freezing Theory was contributed by the Norwegians in the mid-1930s and it will be dealt with first. The change of state, on cooling from invisible water vapour to visible cloud droplets, is enhanced by the abundant presence in the atmosphere of condensation nuclei: tiny particles of dust, smoke or salt (from seaspray). The change from the water state to the ice state demands some similar agency, since water droplets readily super-cool to temperatures as low as minus 40°C. This second agency is provided by freezing nuclei: ice crystals or other

crystals, e.g. silver iodide, which have the same shape.

Ascending cloud-forming air may eventually rise into regions where the temperature is below 0°C (32°F), the region where freezing nuclei exist. These nuclei have a great affinity for water and, in the presence of water droplets, soon grow into snowflakes. Eventually these become too large to be supported by the rising air and they descend. On passing through warmer layers they melt and fall to the ground as rain. If the air is already cold near the ground (temperature less than about 2°C or 36°F) there will be insufficient time for melting and snow will occur at ground level. With near surface temperatures a little higher, say between 2°C (36°F) and 4°C (39°F), some snowflakes will melt and then both rain and snow reach the ground; this is called sleet. This theory obviously accounts satisfactorily for snow and sleet, as it also does for rain, at times, in all latitudes. However, it cannot account for precipitation from cloud whose tops are warmer than 0°C (32°F) and where, obviously, no freezing stage can occur. This happens more frequently in tropical regions and sometimes in the temperate climate belt.

A second theory is therefore required and it is called the

Type of drop	Diameter mm	Terminal velocity m/s
Cloud	0.01	0.003
	0.02	0.012
	0.1	0.27
Drizzle	0.2	0.72
	0.4	1.62
	1.0	4.03
Rain	2.0	6.49
	4.0	8.83
	5.0	9.09
	5.8	9.17
Hail	Up to 75	Up to 50

Table 1. Cloud and precipitation droplet sizes and terminal velocities (maximum fall-out speeds in still air)

Coalescence Theory. This requires that some cloud droplets be larger than others and, therefore, that their rates of ascent will differ. The larger ones will become too heavy and begin to descend, collecting more smaller droplets before falling out of the cloud as rain. The coalescence theory has long been accepted for 'warm' cloud precipitation. In fact, even in 'cold' clouds (producing rain not snow at the ground) it may play a secondary role to the freezing theory.

The intensity of the precipitation, i.e. whether it is classed as slight, moderate or heavy, depends on the rate of ascent of the air in which the parent cloud forms. When the ascent is very gentle then even quite small droplets become too heavy and fall to the ground as drizzle. (In cold conditions its frozen equivalent is known as snow grains.) The parent cloud for drizzle and snow grains need only be a few hundred metres in depth. When the rate of ascent is greater considerably larger drops can be supported before precipitation occurs resulting in moderate or heavy rain associated with nimbostratus and large cumulus or moderate sized cumulonimbus clouds (in all of these cases up to about 2,000 to 5,000m deep).

The strongest up-currents occur in large cumulonimbus clouds and therefore the heaviest rain is associated with them. These clouds usually attain depths of 6,000 to 9,000 m, and sometimes exceed 20,000m in tropical regions. The up-currents in mature 'tropical' cumulonimbus clouds attain considerable rates (perhaps up to 100 knots) and the associated rain is torrential. Moreover, within these clouds the patterns of vertical motion are so complex that vigorous up-currents may exist alongside vigorous down-currents and there may be several such columns of up- and down-currents in the same cumulonimbus cloud. These are the conditions under which hailstones are formed.

Large cloud droplets, on being carried upwards into 'colder' regions (colder than 0°C), are subject to the freezing process. They will eventually descend into 'warmer' regions, growing meantime by collision (coalescence) where melting occurs. Then they may re-enter an up-current and return to the cold part of the cloud where refreezing soon occurs, now as a layer of translucent ice. While in the upper part of the cloud smaller ice particles will adhere on impact, with air trapped between them, before renewed descent occurs with subsequent melting of the outside of the recently acquired 'small particle' layer. This process of ascent and descent may continue several times so that the hailstone is finally composed of alternate layers of opaque and translucent ice. In temperate latitudes (about 40° to 60°N and S) hailstones are typically lentil or pea sized. Exceptionally, in more tropical latitudes they attain the size of a tennis ball.

Cumulonimbus clouds are also the only cloud type producing thunderstorms. The large water drops they contain frequently split into smaller drops in the often vigorous up- and down-currents within these clouds. This leads to charge separation so that the upper part of the cloud becomes positively charged and the lower part negatively charged. Lightning is the visible effect of the discharge between oppositely charged areas, whether from cloud base to top, cloud base to ground, or sometimes cloud top to space. The lightning, due to the ionization of the air along its path, is frequently forked and is called forked lightning. (So-called 'sheet' lightning is simply forked lightning obscured by cloud or rain.) As a point of interest, in the case of cloud to ground discharge an invisible leader stroke, or strokes, often stepped, reaches towards the ground. The lightning normally observed is a very bright and short-lived return stroke from ground to cloud which occurs when the leader stroke touches or almost touches the ground. The ionization causes the affected air to undergo sudden expansion: the shock waves set up by this expansion are heard as thunder.

Though they occur simultaneously at their source, due to the difference between the speed of light and that of sound the thunder is heard after the lightning is seen unless the thunderstorm is overhead. As a guide, the distance in miles from the observer to the thunderstorm may be estimated by dividing the time interval between the lightning and the thunder by 5. In calculating the 'distance off' when the lightning is fairly frequent, it is obviously important to associate the thunder with the correct lightning stroke. Normally, thunder may be heard at distances up to 10 miles, and on rare occasions up to 20 miles. Thunderstorms are more frequent over land but they do occur over the sea, and small and large vessels have sometimes been struck by lightning. Though obviously alarming, there are no reports of any serious injuries to crew. However, navigation equipment e.g. compass, radio, radar etc, will be seriously affected.

One further phenomenon associated only with cumulonimbus cloud is the waterspout (see Plates), a potentially serious hazard to yachts. Waterspouts are more common in tropical latitudes than elsewhere but they are also observed in temperate regions. They are vertical vortices containing a core of low pressure, which reach down from the cloud base 500m or so to the sea surface. Well developed waterspouts exceptionally may have horizontal dimensions of about 100m but are usually only a few metres across. They begin at the cloud base itself and descend as a rotating funnel of cloud. The lower, invisible, 'dry' part of the vortex may be seen to agitate the sea surface and lift spray before the descending funnel of cloud reaches the surface. Sometimes there is no further development and the vortex soon dies away, but at other times the funnel cloud reaches the sea surface and it would then appear that sea water is being transported vertically into the main cumulonimbus cloud. This is not so: the waterspout is composed of cloud droplets, though there may be considerable spray at its base; the darker the appearance the more vigorous the circulation around the waterspout.

There is little or no information on wind speeds around a waterspout vortex but these may well be far in excess of 50 knots in extreme cases. Individual waterspouts (and several may be observed simultaneously from the same cloud) may last up to half an hour or so. Though the movement of a waterspout may be fairly erratic, for it may sometimes be seen to snake about, its overall movement will be roughly the same as that of the parent cloud; typically between 10 and 25 knots. The base of the cumulonimbus cloud (for the upper parts may be obscured by other clouds) should be continuously observed, in the same way that another vessel on a possible collision course is constantly watched, to determine its relative movement for waterspouts are clearly a serious hazard to small craft and are to be avoided if at all possible.

In the event that a yacht is unable to avoid a waterspout, all sails should be handed and securely stowed and all hatches opened so as to diminish the risk of 'explosion' due to the difference between air pressure within the yacht and the much lower pressure in the centre of a mature waterspout vortex. (Though in rough seas some will have to remain closed to avoid taking water below.) Some yachts have experienced waterspouts at very close quarters; some have even sailed deliberately into immature waterspouts. It is sufficient here to say that an immature waterspout may develop into dangerous maturity in a matter of seconds. Waterspouts inevitably decay soon after crossing the coast but they have been known to cause considerable damage, and claim lives, on land in coastal areas beforehand. They are the waterborne equivalent of tornadoes, though fortunately much less vigorous.

CHAPTER 8

Depressions and Fronts

MOST bad weather is associated with depressions since they provide the essential ingredients: rising air for thick cloud and rain, and steep pressure gradients for strong winds. In this chapter we shall examine the distribution of weather and wind around a depression and we shall begin with its associated fronts which separate the warm tropical air from the colder polar air (see Diagrams 10 and 11, stage 3).

For reasons largely connected with friction between the air and the ground surface, and also with the curvature effect described in Chapter 4, the air in the warm sector moves more quickly than the cold polar air ahead of it (on its eastern side). Being less dense, it slowly rises over the cold air at their common boundary, with a typical slope of about 1:150 (Diagram 32a). By definition, since warm air is overtaking cold air, the boundary is known as a Warm Front. It is a region of 'gently' ascending air which eventually produces a broad band of cloud from which continuous rain (snow) often falls for many hours. A cross-section of a warm front is shown in Diagram 39.

In the rear of a depression, the cold air is normally moving more quickly than the warm air ahead of it, on its eastern side. The advancing cold air undercuts the warm air as a wedge and forces the warm air to ascend at their common boundary (Diagram 32b). Now that cold air is overtaking warm air the boundary is known as a Cold Front. (Both Warm and Cold Fronts are names given to particular sections of the Polar Front when within the circulation of a depression.) At a cold front, the forced ascent of warm air is not so gentle and the typical slope here is about 1:50. This leads to a narrower belt of rain, which is sometimes heavier than at a warm front (Diagram 39).

The depression's cold front is normally moving more quickly than its warm front (see Chapter 12). The cold front eventually catches up with the warm front and the warm sector becomes progressively narrower, the process beginning at the centre of the depression and extending outwards (Diagrams 34 and 35), resulting in the warm air being steadily lifted away from the ground surface (Diagram 33). This is known as the Occlusion Process and the resultant front is called an Occlusion. The polar air in the van of a depression is normally less cold than that to its rear, so that as the warm sector is lifted away the two polar air masses are allowed to meet. However, they do not mix, the advancing colder air to the rear undercuts the polar air now ahead of it (Diagram 33c) to form

32a. Vertical section showing tropical air overtaking and rising over polar air at a warm front.

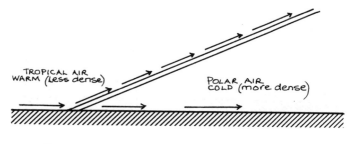

TROPICAL AIR
WARM (less dense)

POLAR AIR
COLD (more dense)

32b. Vertical section showing polar air overtaking and under-cutting tropical air at a cold front

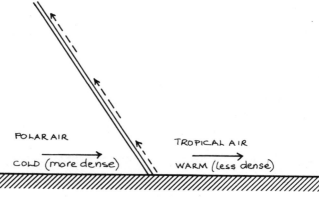

In both diagrams it can be seen that it is the warm (tropical) air which rises; the cloud is therefore mainly contained in the warm air (see Diagram 38).

POLAR AIR

COLD (more dense)

TROPICAL AIR

WARM (less dense)

a Cold Occlusion (Diagram 34); this is the normal case.

Exceptionally, the polar air ahead of the low is the colder air mass. In this case the advancing polar air in the rear of the low, being less dense, rides over the colder air to form a Warm Occlusion (Diagrams 33d. and 35). Vertical sections through both types of occlusion are shown in Diagram 36. In many ways warm occlusions may be regarded as warm fronts and cold occlusions as cold fronts.

Fronts are identified on weather charts by internationally used symbols. A series of solid semicircles indicates a warm front, a series of solid triangles a cold front, and a mixture of the two identifies an occlusion. A cold occlusion follows the line of the original cold front (Diagram 34) and a warm occlusion that of the original warm front (Diagram 35). With all fronts the symbols face in the direction towards which the frontal boundary is moving. These symbols are commonly used by press and TV throughout the world and in public weather displays. Working charts used in UK and some other met offices show a red line for a warm front, blue for a cold, and purple for an occlusion.

To return to the textbook depression, the distribution of cloud and rain around a low is shown in Diagrams 37 and 38. This long-known pattern is entirely confirmed by satellite cloud photographs (Plate 17 and 18). The main mass of cloud and weather is due to the ascent of air within the central vortex; the extensive cloud 'limbs' are due to the ascent of air along the fronts (the narrower cloud belt on the cold front is clearly indicated).

Not all fronts (of any type) produce the same intensity of weather, in the same way that not all depressions are equally vigorous. Some fronts may pass by, especially in summer, with just a broken layer of cloud with little or no wind change; these fronts are either associated with weak or filling depressions, or are themselves a very

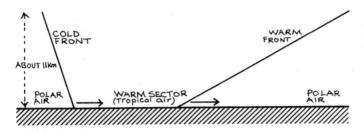

33. Vertical section showing occlusion process: (c) cold occlusion, (d) warm occlusion. The interval (a) to (c) or (d) is about 24 hours

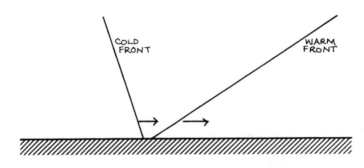

33b. Cold front almost caught up with warm front (at ground level)

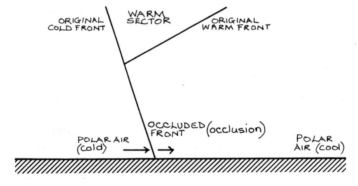

33c. Warm front now overtaken: warm sector lifted as cold polar air runs underneath. Since overtaking air is colder it behaves as a cold front and is called a Cold Occlusion

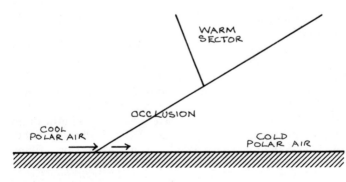

33d. Exceptionally, the polar air ahead of the low is colder than that to the rear. When the latter catches up the resulting occlusion is called a Warm Occlusion

34. The process of occlusion (the occluded front) – northern hemisphere

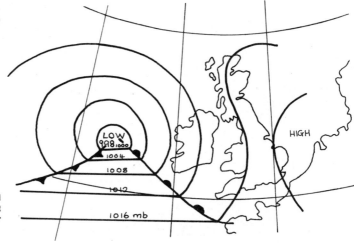

Stage 1
Low moving E. Still deepening. Cold front typically moving faster than the warm front

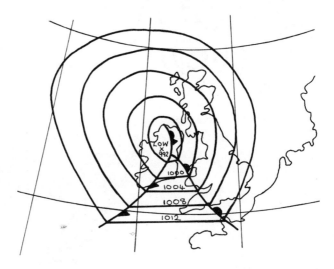

Stage 2 (about 6–12 hr later)
Cold front begins to overtake warm front near centre of low: initial stage of occlusion. Low now normally ceases to deepen; it slows down and begins to turn poleward

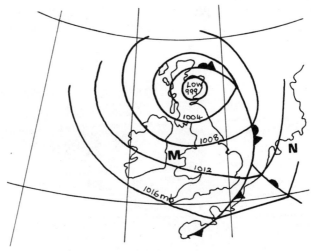

Stage 3 (further 6–12 hr later)
Cold front has overtaken most of warm front: occlusion almost complete. Low begins to fill and becomes very slow moving, sometimes stationary

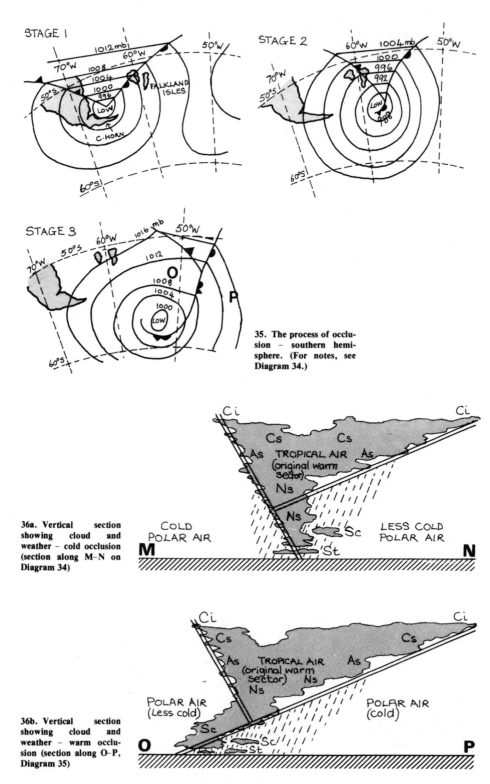

STAGE 1

STAGE 2

STAGE 3

35. The process of occlusion – southern hemisphere. (For notes, see Diagram 34.)

36a. Vertical section showing cloud and weather – cold occlusion (section along M–N on Diagram 34)

36b. Vertical section showing cloud and weather – warm occlusion (section along O–P, Diagram 35)

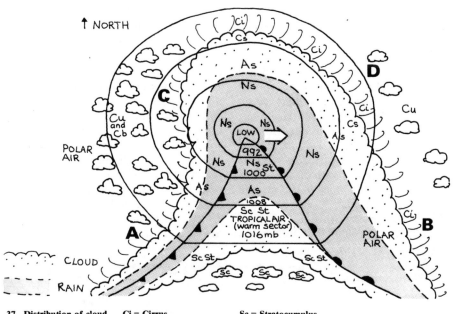

37. Distribution of cloud and rain around a typical depression – northern hemisphere

Ci = Cirrus	Sc = Stratocumulus
Cs = Cirrostratus	St = Stratus
As = Altostratus	Cu = Cumulus
Ns = Nimbostratus	Cb = Cumulonimbus

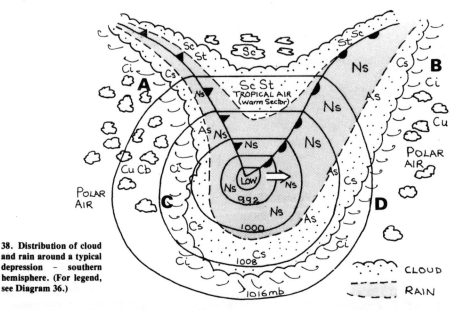

38. Distribution of cloud and rain around a typical depression – southern hemisphere. (For legend, see Diagram 36.)

long way (say 750 miles or so) from their parent depression.

The pattern of winds around depressions is shown in Diagrams 41 and 42 which illustrate the backing or veering (according to hemisphere) of the surface winds from the isobars, due to friction. Wind speed is related to isobar spacing: the closer the isobars the stronger the wind. As we have already seen (Chapter 4), the curvature of the isobars also effects wind speed.

In a depression, in either hemisphere, wind speeds are considerably reduced if the curvature is well marked. The strongest consistent winds, then, are theoretically to be found in the warm sector (straight isobars) near to the centre (isobars closer together) of an almost circular depression. However, temporary gusts or squalls on the passage of the fronts, especially cold fronts, may exceed the 'steady' wind of the associated warm sector.

In many depressions, however, the isobars are not almost circular. We shall see in Chapter 10 that, typically, the isobars are V-shaped at fronts (the V pointing away from the centre of the low), especially at cold fronts. This often results in almost straight isobars running down towards the cold front in the rear of a depression (Diagram 47). In practice, the strongest winds around a low are normally those behind the cold front (north to northwest in the northern hemisphere and south to southwest in the southern hemisphere).

This fact should always be remembered when considering the course to be taken on the approach of a depression when in the vicinity of land. A safe 'weather' shore ahead of a depression will probably become a

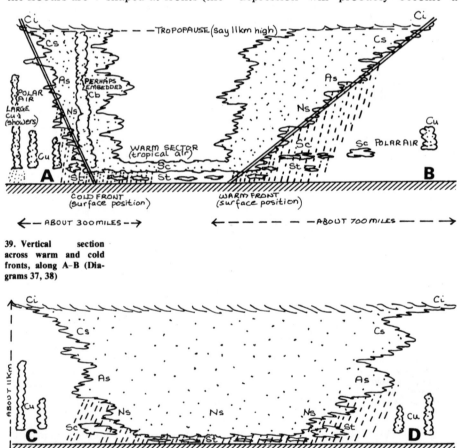

39. Vertical section across warm and cold fronts, along A–B (Diagrams 37, 38)

40. Vertical section on poleward side of depression, along C–D (Diagrams 37, 38)

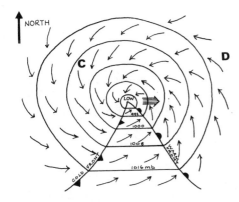

41. Surface winds around a depression – northern hemisphere

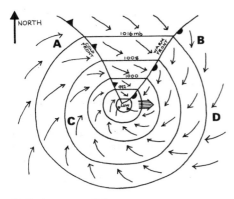

42. Surface winds around a depression – southern hemisphere

dangerous 'lee' shore after the centre of the low, or its cold front, has passed by.

Sharp changes of wind direction also occur at fronts and these are hazardous to small vessels. These sudden changes are often about 90° and may sometimes exceed this especially at cold fronts. The winds shown in Diagrams 41 and 42 are seen to be up to 90° backed from the isobars ahead of the centre and almost parallel to them in the rear. These differences from the 30° or so due to friction (Chapter 4) are considerable and it is as well to pause and explain them since the change of wind, particularly in direction, gives a great deal of information about the movement of a nearby depression. The surface winds just ahead of a moving depression are approximately easterly and those close in on the rear are almost northerly (southerly in the southern hemisphere). This is due to the Isallobaric Effect.

An Isallobar is a line joining places of equal rate of change of pressure or 'barometric tendency'. (The latter is measured, in meteorological practice, every three hours.)

Isallobars define areas of greatest fall or rise of pressure (Diagram 43) which are known respectively as

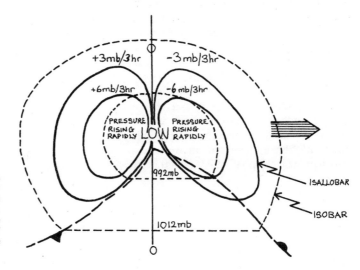

43. Pressure tendencies around a depression – isallobaric highs and lows. Areas of maximum falling pressure are isallobaric lows. Areas of maximum rising pressure are isallobaric highs.

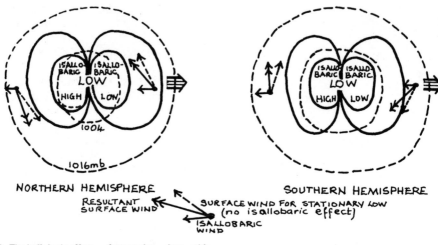

NORTHERN HEMISPHERE   SOUTHERN HEMISPHERE

RESULTANT SURFACE WIND
SURFACE WIND FOR STATIONARY LOW (no isallobaric effect)
ISALLOBARIC WIND

44. The isallobaric effect on the surface wind. NB: the greater the fall or rise of pressure, the greater the isallobaric wind. In a fast-moving low this effect will lead to the winds ahead blowing directly across the isobars.

Isallobaric Lows and Highs. (Depressions tend to move away from the areas of rising pressure towards areas of falling pressure.) Notwithstanding the arguments of Chapter 4, which describes the 'balanced' state of the surface wind under the effects of pressure, friction and the earth's rotation, air particles also move towards areas of falling pressure and away from rising pressure. The results of this effect are illustrated in Diagram 44. The effect, of course, is most noticeable just ahead and to the rear of a moving depression (and to a lesser extent its fronts). There is no effect if a depression is stationary and not deepening since the pressure changes everywhere around it are negligible. It is most marked just ahead of a deepening, fast moving depression (over 30 knots) where surface winds may be backed (veered in southern hemisphere) almost 90° from the isobars.

Since we have just examined the distribution and causes of weather and wind around a depression, we will now examine the signs of the changes of wind and weather as a depression moves towards your sailing area.

CHAPTER 9

Signs of Approaching Bad Weather

AS a depression approaches, pressure will fall, winds will freshen and probably change direction and the sky will become overcast. A careful monitoring of these three elements, pressure, wind and 'state of sky', will normally give advance warning of the approach of bad weather. To illustrate this we shall imagine that we are aboard a well found ocean-going sloop bound from the Solent around Ushant to the Mediterranean. We shall also have to imagine that the yacht's radio has failed and her crew are unable to receive shipping forecasts.

It is mid-day on Saturday 8 September and our yacht is off the Brittany coast about 100 miles east-northeast of Ushant. The log may read 'WNW'ly Force 2, 1018 mb and steady (over the past 3 hours). Weather fine, small amounts of shallow cumulus cloud. Long swell from the westward' (perhaps an indication of a depression in the offing but by no means uncommon in the western Channel). The yacht is on starboard tack carrying full main and no. 1 genoa.

At 1500 hours, feathery wisps of white cirrus cloud are observed in the west and these gradually spread eastwards; meantime the cumulus cloud has moved away towards the eastern horizon. By now the barometer may have dropped a millibar or so and the watch on deck may well be discussing whether there has been a slight backing of the wind towards the west. The watch changes at 1500 hours and the skipper, now on deck, notes these small changes and considers the possibility of an approaching depression. (These signs frequently occur without any further deterioration in the weather, and so far this has been the case.)

By 1800 a milky sheet of cirrostratus cloud has spread over the whole sky. Earlier a halo was observed, but it is no longer visible as the sky in the west is beginning to take on a grey colour indicating the approach of altostratus cloud through which the low sun is seen as an indefinite bright patch. The barometer has dropped a further 3 mb and the wind has now definitely backed to south-southwesterly and increased to Force 3/4. It has been decided that a depression is approaching from some westerly point, possibly quickly, and that it may pass by before the yacht has cleared Ushant so that the strong 'northerly' winds in the rear of the depression would make the Brittany coast a potentially dangerous lee shore. (By this time a yacht holiday cruising in the area, perhaps short-handed, would find it prudent to run for a safe Brittany port, but our yacht and crew are used to strong winds and rough seas and it has been decided to fight this one out in the open Channel).

At 1800, then, course is altered to the northwest to gain sea-room from the Brittany coast, under working jib only. (We are 80 miles southeast of the Lizard and do not wish to be driven too close to the Cornish coast.)

The log extract at midnight reads 'SSE 6 to 7, rain; moderate visibility; sea rough (southerly seas on long westerly swell). Barometer 1006 mb, now falling more quickly having fallen 5 mb in the past 3 hours.' This is a clear indication of stronger winds to come, (see Chapter 15) so the storm jib is set instead of the working jib, mainly to reduce speed.

By 0300 hours the wind is southeasterly Force 7/8 and pressure is falling even more quickly; the rain has become heavier and the westerly swell noticeably bigger. At 0450 the wind suddenly veers to west-southwesterly Force 8; the barometer reads 988 mb. The depression's warm front has passed by. In the following hour the rain becomes much lighter and the barometer has steadied at 988 mb, confirming that we are now in the warm sector and that the centre of the low is passing to the north (Diagrams 41 and 37). The storm jib is handed and the yacht now lies a-hull in mid-Channel. Dawn is breaking revealing a rough sea, moderate visibility with drizzle at times, and the sky is overcast with low stratus clouds which even from the gyrating cockpit are observed to race away to the east. After breakfast (the ubiquitous, but always acceptable, toasted omelette sandwich!) has restored flagging mental agility, there is some discussion on the probable timing of the cold front passage (at this time there are no hints from the weather, wind or

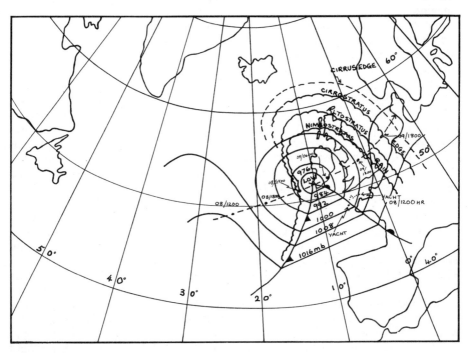

45. Depression and yacht tracks: a severe storm over the Southwest   Approaches and western English Channel

pressure) but the skipper decides to remain a-hull until the cold front has passed when the north or northwesterly winds, providing they are not of storm force, will allow some sail to be carried and a course re-set to clear Ushant.

At about 1045 hours the sky appears a darker grey in the west, the wind seems to have backed and freshened a little. Suddenly, at 1105, the wind veers to north-northwest and increases to Force 9 for a time and it is raining heavily. This is a 'squall' occurring as the cold front passes by. By 1200 hours the sky is clearing dramatically in the west, the barometer has risen 4 mb since 1105, the wind is north-northwest Force 7 to 8 and the rain is dying out. Storm jib and deep-reefed main are hoisted and a course set to clear Ushant. At 2100 the log reads 'WNW 3/4, 1016 mb. Scattered showers, Ushant abeam 8 miles.'

Diagram 45 shows the depression at 0300 on the 9th together with its track from 1200 on 8th to 1800 on 9th, and the track of the yacht, while Table 2 lists the main weather extracts from the log. As an exercise, the reader may wish to make a tracing from Diagram 45 of the isobars, fronts and cloud edges and to slide this along the 'storm' track on Diagram 45 in order to follow Table 2 and the earlier description more closely. As the storm track is slightly curved the warm sector isobars on the tracing should be kept 'parallel' to the track at any given position.

In our example the depression clearly passed by to the north. What if it had passed to the south or overhead?

By reference to Diagrams 37 and 41 and imagining the depression passing to the south (so that the vessel's track would lie along D to C on these diagrams), pressure will fall ahead of the low (except that there is no warm sector and therefore no 'steadying' of the barometer) and the weather sequence be much the same (Diagram 40) but the wind changes will be

Time	Wind	Pressure	Weather	Visibility	Sea
1200	WNW 2	1018 mb steady	Fine	Good	Long W'ly swell
1500	W 2	1017 mb	Ci cloud spreading from W	Good	Long W'ly swell
1800	SSW 3/4	1014 mb	Cs, with As to W	Good	Confused; S'ly seas over W'ly swell
			As to west		
2100			Rain		
2400	SSE 6/7	1006 mb falling even more quickly	Rain	Moderate	Rough and confused
0450	WSW 8	988	Rain→drizzle	Moderate to poor	Rough and confused
0600	WSW 7/8	988	Drizzle	Moderate to poor	Rough
1105	NNW 8/9	986	Heavy rain	Moderate to poor	Rough, confused
1200	NNW 7/8	990	Rain easing. Sky clearing to W	Bec. good	Rough
2100	WNW 3/4	1016	Sct. showers	Good	Moderate to rough

Table 2. Log weather extracts, western Channel: passage of depression to north.

NORTHERN HEMISPHERE

Low passing to north	Low passing overhead	Low passing to south
Wind backs to within a point of SE (and later veers at fronts)	Wind backs to within a point of E and remains there	Wind backs beyond E to NE and N. If process fairly rapid then close pass. If process gradual then low passes at some distance.

SEQUENCE OF WEATHER EVENTS WITH APPROACH AND PASSAGE OF WARM AND COLD FRONTS

Element	Approach of warm front	Passage of warm front	Warm sector	Approach of cold front	Passage of cold front	To rear of cold front
Wind	Backs to S or SE and freshens (perhaps to gale force)	Veers to SW or WSW, and remains steady	Steady SW or WSW'ly wind	Perhaps backs a point and freshens a little	Sudden veer to NW to N. Squally	Slow decrease in speed and slow backing. Wind is now gusty
Barometer	Falling, latterly more quickly	Steadies or falls much less quickly	Steady or slow fall	Falling	Fall suddenly ceases and rapid rise sets in	Rising, progressively less quickly
Visibility	Good, becoming moderate to poor	Moderate to poor	Moderate to poor	Moderate to poor	Rapidly becoming good	Good but moderate in showers
			Perhaps sea fog in some seasons			
Cloud and weather	Cirrus Cirrostratus Altostratus Nimbostratus and rain	Nimbostratus low Stratus Heavy rain	Low Stratus Drizzle (Sea fog?)	Nimbostratus and Stratus Rain	Nimbostratus (perhaps Cumulonimbus) Heavy rain Squally	Soon breaks to Cumulus and perhaps Cumulonimbus - scattered showers and sunny periods

Table 3. Wind changes indicating relative track of passing depression: wind and weather changes at fronts: northern hemisphere (low moving west to east)

SOUTHERN HEMISPHERE

Low passing to south	Low passing overhead	Low passing to north
Wind veers to NE and remains there (later backing at fronts)	Wind veers to within a point of E	Wind veers beyond E to SE and later S

SEQUENCE OF WEATHER EVENTS WITH APPROACH AND PASSAGE OF WARM AND COLD FRONTS

Element	Approach of warm front	Passage of warm front	Warm sector	Approach of cold front	Passage of cold front	To rear of cold front
Wind	Wind veers to NE and freshens (perhaps to gale force)	Wind backs to WNW	WNW steady	Slight veer and slight freshening	Backs to SW or SSW Squally	SW later veering WSW, slow decrease but wind is gusty
Barometer	Falling, latterly more quickly	Steadies or falls much more slowly	Steady or slow fall	Falling	Fall suddenly ceases and rapid rise sets in	Rising, progressively less quickly
Visibility	Good, becoming moderate to poor	Moderate to poor	Moderate to poor. Perhaps sea fog in some seasons	Moderate to poor	Rapidly becoming good	Good, but moderate in showers
Cloud and weather	Cirrus Cirrostratus Altostratus Nimbostratus and rain	Nimbostratus Low Stratus Heavy rain	Low Stratus Drizzle (Sea fog)	Nimbostratus Low Stratus Rain	Nimbostratus (perhaps Cumulonimbus) Heavy rain	Soon breaks to Cumulus and perhaps Cumulonumbus. Scattered showers and sunny periods

Table 4. Wind changes indicating relative track of passing depression: wind and weather changes at fronts: southern hemisphere (low moving west to east)

different. From Diagram 41, along the line D to C (the relative track of the vessel) the wind begins in the southeast and gradually backs to northeast and later northwest. During the backing process it will also freshen.

If the low were to pass overhead, the wind would back from its earlier southeasterly direction towards easterly and remain there increasing in strength meantime, until the low passed overhead when, perhaps after a short lull, it would suddenly pick up from a northerly point.

In either case it can be seen that standing out away from the Brittany coast is a safe action.

From the foregoing, the following early guidance may be deduced:

If, in the northern hemisphere, the wind backs to within a point or so of southeast and no further, the low will pass to the *north*. (We are here discussing early guidance: of course the wind will later veer with the passage of the fronts.)

If the wind backs to within a point or so of east and remains there, then the centre of the low will pass *overhead* in the vicinity of the yacht.

If the wind soon backs beyond east and continues to back, the low will pass to the *south* (see Diagram 41). In the latter case, if the backing is fairly rapid the low will pass close southward, and if it is gradual it will pass some distance to the southward (close implies gales, whereas some distance implies less strong winds).

In the southern hemisphere the approach of a depression is signalled in much the same way (see Diagrams 38 and 42). The cloud and pressure sequences are typically identical and the wind backs as a depression passes on the poleward side (vessel's relative track B - A in Diagram 42) and veers as it passes on the equatorial side of the vessel (D - C in Diagram 42).

A similar set of rules apply in this hemisphere, as follows:

If, in the southern hemisphere, the wind veers to within a point or so of northeast and no further, the low will pass to the *south*. (Again, it will of course back later on at the passage of the fronts.)

If the wind veers to within a point or so of east and remains there, the low will pass *overhead*.

If the wind veers beyond east to southeast and later south, the low is passing to the *north*. If this latter process is fairly rapid, the low will pass close north.

These 'rules', which will be found to work on nearly all occasions (but lows do sometimes turn unpredictably), should enable one to sort out the weather/lee shore situation both ahead of and in the rear of a low when in the vicinity of land, and give the ocean skipper an indication of which course to take to avoid the strongest headwinds and seas. They are listed in Tables 3 and 4 together with the changes across fronts. (Tropical revolving storms, typhoons etc, are different and are given separate treatment in Chapter 17.)

On a completely different time scale, now a matter of minutes, the 'squall' presents signs of 'approaching bad weather' but the associated strong winds and rain last for only a few minutes. Since a squall is essentially a wind effect its description will be deferred to Chapter 12.

It will do no harm here to repeat the message of the first two sentences of this chapter: falling pressure, thickening state of sky, backing (veering in southern hemisphere) and freshening of the surface wind, particularly if sustained over more than one to two hours, invariably mean the approach of bad weather which will last for at least a few hours. These three elements should be continuously monitored.

CHAPTER 10

Other Pressure Features and Associated Weather

A typical weather chart, displaying lows (depressions), highs (anticyclones), troughs, ridges and cols, is shown in Diagram 46. Lows and their associated weather have already been discussed in Chapters 8 and 9.

Highs, being areas of descending air, are areas of fine weather. Towards the centre of the high, and this may cover many hundreds of square miles, winds are light and variable. However, around the periphery, winds can sometimes be quite strong, Force 6 or more, especially if a depression is attempting to move towards the high pressure area (see also end Chapter 11). High pressure areas are hardly ever circular in shape; they are frequently elliptical and sometimes more irregular. The 'extension' of an anticyclone along the major axis of the ellipse is called a Ridge of high pressure. Ridges are also areas of fine weather (descending air) and of light winds, but once again strong winds may be experienced on one side or other of the ridge.

Lows, too, are rarely circular in shape. They are deformed by Troughs of low pressure radiating outwards from their centres. Along the axis of the trough the pressure is lower than at other points on either side of the axis equidistant from the centre of the low (Diagram 47). In other words a trough is a V-shaped bend in the isobars pointing away from the centre of the low. Isobars are invariably troughed at fronts; these are called Frontal Troughs and their associated weather has been described in Chapter 8. Other troughs are not infrequently found within the circulation of a depression. Such a non-frontal trough is shown in Diagram 47 in the cold air to the rear of the depression. Whereas on either side of this non-frontal trough scattered showers may be occurring in the polar air mass, heavier and more prolonged showers will be occurring within the trough. This is due to the fact that relatively sharp curvature of the isobars leads to increased ascent of air; hence the more vigorous convection on the trough.

Troughing may be so vigorous in winter, especially in Arctic air masses arriving in more temperate latitudes, that a small but quite violent depression forms in the trough; this is called a Polar Low. It always remains subsidiary to the parent low and usually decays when it finally swings round into the 'equatorward sector' of the parent low (Diagram 48).

Cols are the 'slack', flat areas between pairs of anticyclones and pairs of depressions. Since they are regions without pressure gradients they are windless areas. Overland fog may form

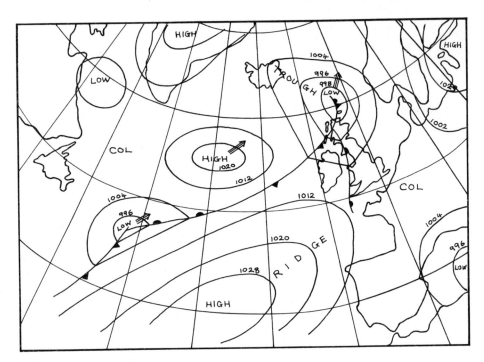

46. Typical weather chart showing the various pressure features

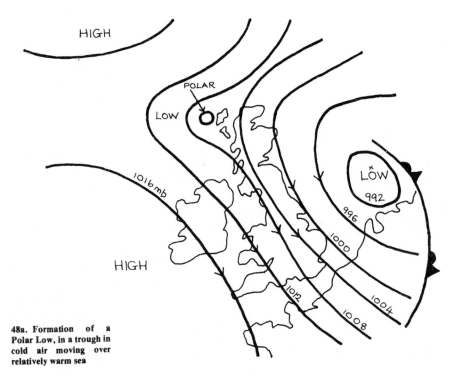

48a. Formation of a Polar Low, in a trough in cold air moving over relatively warm sea

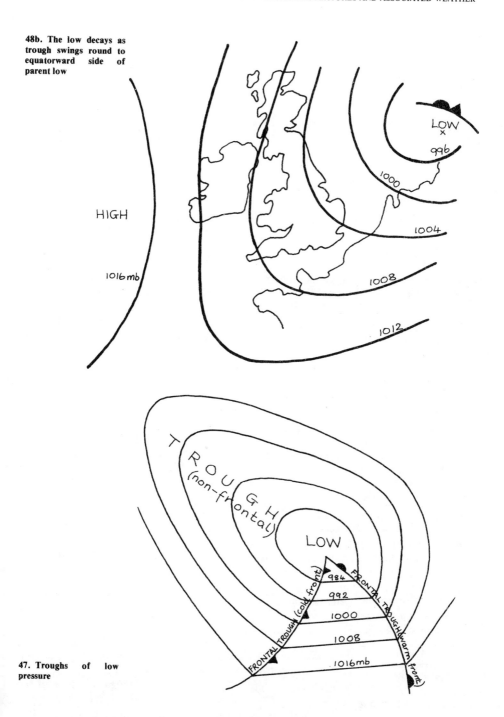

48b. The low decays as trough swings round to equatorward side of parent low

LOW
×
996

1000

1004

1008

1012

HIGH

1016 mb

T R O U G H
(non-frontal)

LOW

FRONTAL TROUGH (cold front)

FRONTAL TROUGH (warm front)

984

992

1000

1008

1016 mb

47. Troughs of low pressure

under a col by night and this may spread into coastal waters as it clears next morning. In summer thunderstorms may occasionally form in cols, otherwise they are of no particular note to yachtsmen.

CHAPTER 11

Movement of Pressure Features

WAVE depressions sometimes travel at 60 knots. This means that a low may be wreaking havoc in the Southwest Approaches and Western Channel in a little over 24 hours after clearing Newfoundland! Equally spectacular in its own way, the centre of an anticyclone may appear to move hundreds of miles, completely altering the wind pattern over thousands of square miles in a day or even less.

These dramatic examples serve to illustrate that weather systems move – sometimes rapidly. This chapter describes the movements of the various pressure features and, where possible, accounts for them, but it is fair to state now that most of the information predicting their movement is not going to be available to the yachtsman or mariner at sea. He will have to rely on the rules of thumb already given in Chapter 9 together with other indications which will follow in this chapter.

Where shipping forecasts are available there is no substitute for taking note of each broadcast, but there are of course large areas of the oceans where no such guidance is provided. Some background knowledge of the normal behavioural patterns of pressure features would therefore be beneficial; this to be supplemented by whatever indicators present themselves in each individual case.

Polar Front depressions, forming as they do in the westerly wind belts of temperate latitudes, normally move from west to east, or more typically from west-southwest to east-northeast, through this belt. Occasionally, however, the westerly wind belt (normally a low pressure area) may be interrupted in a given area for a few days or even a week or two by the sub-tropical high moving poleward into temperate latitudes. This high then becomes a substantial barrier to the eastward movement of depressions which are then 'steered' around the high. The high has to recede into its normal tropical latitudes before the usual eastward progression of lows resumes. Another rule about the movement of lows is that, while its fronts are not occluded, a low moves parallel to the warm sector isobars at about two-thirds of the geostrophic wind measured from these isobars. At first sight this may seem of little value to a yacht at sea, but if the yacht is in the warm section her crew can make a ready estimate of the speed of the low from the wind they experienced. The surface wind in the warm sector is itself about two-thirds of the geostrophic wind so it will immediately give an

68

indication of the speed of the low. The direction of the surface wind in the warm sector is about 25° backed (veered in the southern hemisphere) from the isobars; the appropriate adjustment (for hemisphere) made to the warm sector surface wind direction will therefore give a good indication of the direction of movement of the low.

Since some lows behave in oblivious ignorance of this 'warm sector' rule, a close watch should be kept on the barometer for a continued fall of say 2 to 3 mb/hour in the warm sector (normally little or no fall here), indicating that the low is stil deepening or is moving towards the yacht.

As another general rule, once the occlusion process has started, a depression begins to slow down, does not deepen any further, and turns progressively poleward (Diagrams 34 and 35). By the time occlusion is almost complete the low is normally almost stationary and filling. Should another wave depression approach the area it will usually absorb the old low in its own circulation where it quickly disappears. On some occasions, however, the original low maintains its identity for some time (several days) after reaching full maturity. Instead of becoming absorbed by the next

depression, the two almost equal in importance, rotate dumb-bell fashion about a common axis for a day or two before one or other decays.

Fronts revolve around the associated low centre. The warm front moves at about two-thirds of the geostrophic wind measured along the front; a cold front at full geostrophic speed measured along itself. This higher fraction, together with the fact that the geostrophic wind along the cold front is usually greater than that along the warm front, ensures that the cold front invariably catches up with the warm front and occlusion results. Fronts obviously revolve in an anticlockwise direction around northern hemisphere lows and in the opposite direction in the souther hemisphere.

Non-frontal troughs also revolve about the associated low trough in a less predictable fashion than frontal troughs. In any case they rarely last for more than a day or so.

The movement of highs is an entirely different consideration and here it is necessary to differentiate between different kinds of anticyclones. The semipermanent anticyclones such as the Azores, Bermuda, Hawaiian Highs and those of the southern hemisphere oceans are called 'warm' highs. It is perhaps easy to remember this since

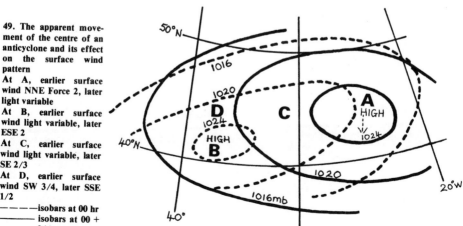

49. The apparent movement of the centre of an anticyclone and its effect on the surface wind pattern
At A, earlier surface wind NNE Force 2, later light variable
At B, earlier surface wind light variable, later ESE 2
At C, earlier surface wind light variable, later SE 2/3
At D, earlier surface wind SW 3/4, later SSE 1/2

– – – –isobars at 00 hr
——— isobars at 00 + 24 hr

they are themselves normally found in warm tropical latitudes. Warm anticyclones are often almost stationary for weeks at a time and when they do move they seem to jump incredible distances in a very short time. This is due to the fact that their maintenance or decline is controlled by conditions in the upper troposphere which themselves may change dramatically in a very short time. The situation here is sometimes favourable for the decline of one part of an anticyclone and for the intensification of another possibly in a day or less, so that the centre of the anticyclone seems to have moved perhaps several hundred miles in that time. It is of course only an 'apparent' movement. However, in this time the anticyclone may entirely change its shape with considerable changes to the pattern of winds around it (see Diagram 49).

Another type of high is known as a 'cold' anticyclone. This is found on the cold (poleward) side of the Polar Front and is an area of higher pressure between two travelling depressions (Diagram 46). It moves in much the same way as its escorting depressions. Cold anticyclones are sometimes accompanied by two deep travelling depressions. Strong, or even gale force, winds may then occur over the greater part of the anticyclone except, perhaps, for the area within 100 miles or so of its centre. Such 'windy highs' are more likely in winter, when the escorting depressions are normally more vigorous than in summer.

There are no rules concerning the movement of cols. Their existence is due to the juxtaposition of highs and lows. While they remain stationary the col persists. More typically, the pattern is constantly changing as lows fill and are replaced by others and perhaps the highs intensify and drift a little one way or another so that the geometry of the col is constantly changing.

CHAPTER 12

Wind

S TRONG winds and their constant companions, rough seas, probably present the most serious hazard faced by yachtsmen. The wind is, also, sailing yachts' main means of propulsion. For these reasons it can be argued that yachtsmen should be as familiar with this subject as they are with navigational matters and also the workings of their auxiliary engines. This chapter will be entirely devoted to the wind and its variability, its gustiness, its local differences near coasts, its diurnal differences and its regional peculiarities.

The wind is simply the movement of air occurring due to differences in surface atmospheric pressure from one place to another. A study of the wind of merely a few minutes soon reveals its major characteristic, its variability. In fact, in the lowest few metres of the atmosphere where *homo sapiens* lives and sails his yachts, the wind is never steady; it is constantly changing in both direction and speed.

This is as true in the longer term, say over a few days, as it is over a period of a few minutes. Even the trade winds, the steadiest winds of all, show changes from day to day. In the temperate latitude belts of the disturbed westerlies, which show more variability than in any other belt, the changes in speed and direction are due to the movement of large-scale eddies in the atmosphere, depressions and anticyclones. The minute to minute variability of the wind is also due to eddies but these are of a much smaller order of scale. Turbulent eddies, perhaps only a few metres across, constantly develop and decay in the atmosphere especially in its lower layers and their movements are the cause of the short-term variability in the surface wind. Friction over a rough land surface increases the turbulence and so as a general rule the wind shows more variability over land than over the open sea. Diagram 50 shows a typical wind record, or anemogram: it can be seen that fluctuations in direction of about 20° on either side of the mean direction and variations of about 25 per cent from the mean speed constantly occur every half minute or so. Over rough land there are much larger variations in speed and direction sometimes over a period of only a few seconds.

The peaks of increased wind speed are known as Gusts and the troughs as Lulls. The percentage ratio of the increases in the gusts to the mean speed is the Gust Factor, typically about 50 per cent over the open sea. The gust factor is much higher on some days than on others, depending on the stability of the air mass affecting the

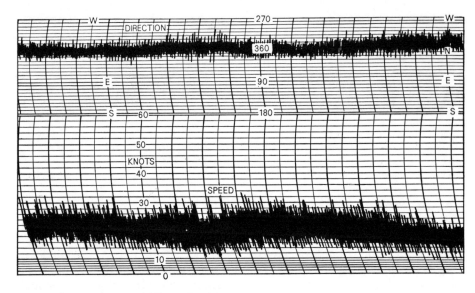

50. A typical anemo-graph trace showing variability of surface wind direction and speed

area. It is best developed in an unstable air mass, i.e. one that encourages vertical motion of the air. Cumulus clouds, characteristic of an unstable air mass, result from heated air rising from the ground surface in the form of 'bubbles'. This air is replaced by air from about 1 km above the surface - the level of the geostrophic wind, which prevails above the friction layer next to the ground. The descending air retains its direction and speed and arrives at the surface as an intermittent gust, shown schematically in Diagram 51. The gust is, then, not only a temporary increase in wind speed but also a change in direction. In the northern hemisphere the direction veers (about 20° to 30°) in gusts and in the southern hemisphere it backs by the same amount. From these facts the following rule is derived:

When beating to windward in gusting conditions, all other things being equal, starboard tack is the favoured tack in the northern hemisphere and port in the southern, since gusts give a lift on these respective tacks. (See Diagram 51.)

A minute or two's notice of the arrival of gusts may be obtained by careful observation to windward where gusts may be seen as approaching localized disturbances of the sea surface, such as catspaws. It may be noted, too, that gusts are more frequent and stronger when one is close to cumulus clouds. It is also true that on many occasions in otherwise calm conditions some wind may be found in the close vicinity of cumulus clouds due to the vertical air motion producing the cloud. In all these cases the gust is caused by the upper level (geostrophic wind) descending to the surface in exchange for air which has been convected upwards through surface heating. For the purposes of this book they will be called 'air mass gusts'.

Quite a different and, since it is more violent and less predictable, more hazardous gust occurs in association with showers and thunderstorms (from cumulonimbus cloud). These gusts are due, at least in part, to the downrush of air associated with the falling raindrops and will be referred to as

51. Wind changes in gusts

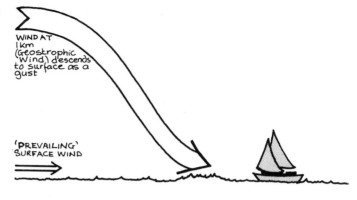

a. A gust is the upper level (geostrophic) wind descending to the surface

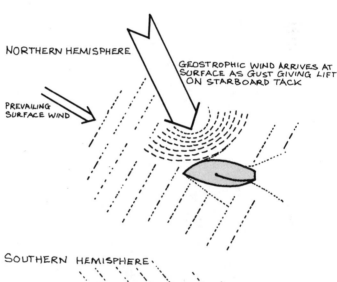

b. Wind direction *veers* in gusts in the northern hemisphere

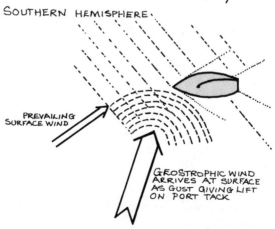

c. Wind *backs* in gusts in southern hemisphere

'downdraft gusts'. Such a gust, if prolonged for more than a minute, is designated a Squall. The downdraft of air beats down on the surface and then fans out radially in the semicircle into which the rain-storm is moving: there is little downdraft effect to the rear of the storm (see Diagram 52). The

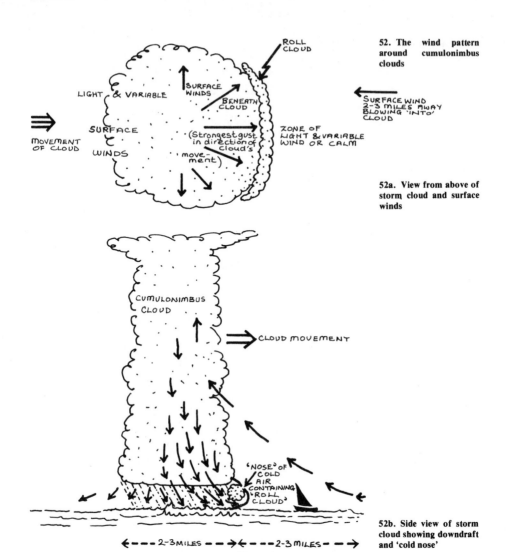

52. The wind pattern around cumulonimbus clouds

ROLL CLOUD

LIGHT & VARIABLE

SURFACE WINDS

BENEATH CLOUD

SURFACE WINDS

(Strongest gust in direction of cloud's movement)

MOVEMENT OF CLOUD

ZONE OF LIGHT & VARIABLE WIND OR CALM

SURFACE WIND 2-3 MILES AWAY BLOWING 'INTO' CLOUD

52a. View from above of storm cloud and surface winds

CUMULONIMBUS CLOUD

CLOUD MOVEMENT

'NOSE' OF COLD AIR CONTAINING 'ROLL CLOUD'

52b. Side view of storm cloud showing downdraft and 'cold nose'

◄----2-3 MILES----►◄----2-3 MILES----►

maximum strength of this type of gust or squall is found directly ahead of the parent cloud.

Since atmospheric pressure changes little with the approach of strong gusts and downdraft squalls, the barometer will give no indication of their approach.

To avoid these squalls it is necessary to know the track of the cumulonimbus cloud. This itself will be steered by the wind meaned over the whole depth of the cloud - at least several kilometres - which is frequently very different from the surface wind.

To determine the track there is no substitute for frequent bearings of the middle of the shower, which will be observed as a grey fuzziness beneath a dark cloud base against an otherwise fairly clear horizon. The surface wind itself is not a good indicator as we shall soon see.

A well-developed cumulonimbus cloud, capable of producing heavy showers or thunderstorms, affects the surface winds in a wide area in its vicinity perhaps up to 5 miles radius and sometimes more from the edge of the shower, which itself may be 5 to 10

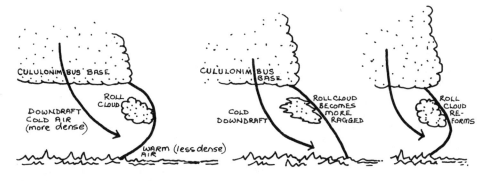

53. The breakdown and reformation of the downdraft cold nose in cumulonimbus clouds

Stage 1
Friction at surface retards progress of cold downdraft: cold nose develops with cold air overlying warm air: 'unstable' condition.

Stage 2
More dense cold air in 'nose' rushes down to surface, destroying nose. Roll cloud becomes more ragged.

Stage 3
Friction again slows down progress of cold air: 'nose' and roll cloud re-form.

miles across.

Diagram 52 gives an idealized representation of the surface wind pattern around a cumulonimbus cloud both in section and in plan. Well ahead of the rain-storm, the surface wind blows gently towards the storm. Within a mile or so of the edge of the storm this air rises away from the surface and eventually enters the storm cloud. There is then usually a zone of light and variable winds until the arrival of the storm cloud. Meantime the downdraft from within the storm has formed a 'nose' of cold air just ahead of the storm edge, by friction at the ground surface (Diagram 53). The nose of cold air is an unstable condition in that it contains cold (denser) air which is overlying less dense air and which constantly strives to descend to the surface. It repeatedly succeeds, breaking down the nose as it does so, only for it to be reformed by friction as the cold air continues to advance (Diagram 53). The breakdown of the cold nose adds to the effect of the squall.

A dark, arched roll of cloud, often visibly turbulent, develops within the nose of cold air. This is often a very distinctive threatening feature which appears to race overhead on its final approach. Its arrival overhead coincides with the onset of the gust or squall. The wind may suddenly rise from near calm to over 40 knots in a few seconds. The rain, often heavy, will follow almost immediately. The gust or squall soon dies down to around half its original speed within a few minutes, and later as the rain becomes less heavy the wind becomes quite light.

The first sign of the approach of this type of gust is the roll-cloud; the second is the state of the sea underneath the roll-cloud. The latter is a good indication of the violence of the gust to come. However, the changed nature of the sea surface can normally only be observed when the storm is a mile away or less. The speed of movement of the whole system is typically 25 knots (the range is probably from about 10 to perhaps 50 knots) so that at, say, half a mile there is normally a little over a minute before the squall hits the yacht, which will be just sufficient to reduce sail or even hand all sails.

Though all roll-clouds when associated with heavy showers are clear harbingers of squalls, some roll-clouds which still take on a dark, threatening appearance develop in front of clouds which are not producing (and never will produce) showers. As they arrive overhead there is little or no variation in what is already a light wind. This non-starter is very difficult to pick out in advance, though the sea state underneath it will often give a clue: the only safe thing to do is to assume it will develop a blow and at least reduce sail.

Squalls of a similar nature often occur at active cold fronts, but here the sudden increase of speed may well be accompanied by a veer (northern hemisphere) of about 90° in the wind's direction. Diagram 54 shows the position of the cold nose on the frontal boundary: its temporary breakdown followed by its redevelopment is similar to that shown in Diagram 53. The breakdown of the cold nose at the cold front enhances the gust (squall) effect as the front arrives. Incidentally, it is the leap-frogging effect of the repeated breakdown of the cold nose which chiefly accounts for a cold front moving faster than the warm front ahead of it.

In some parts of the world cold fronts sweep across the area in advance of well-known regional winds. For example the Norther of the Gulf of Mexico is the northerly behind a cold front; this is usually a winter phenomenon. The Southerly Buster on the east coast of Australia is another post-cold front wind, as is the Pamperro off the east coast of South America which, perhaps, is the most notorious squall, sometimes exceeding 70 knots.

Cold front squalls take on the appearance of a line - the line of the cold front. Such lines are known as Line Squalls. The line is perpendicular to the direction of movement. For those yachts fitted with 3 cm radar, line squalls (as well as individual showers or thunderstorms) are clearly visible on radar displays.

Line squalls also occur independently of cold fronts. Though they may be found almost anywhere, they are more frequent in non-frontal troughs in temperate latitudes, and more especially in the Doldrums belt. In fact in the Doldrum zone most bad weather is 'organized' into line squalls, or Convergence Zones. Two regularly occurring squalls in this latitude belt are worthy of mention. The Sumatra is a line of showers or thunderstorms orientated north/south and moving eastwards across Sumatra and the Malacca Straits to West Malaysia, Singapore and even further east. Its arrival is typically announced by a squall of 30 to 40 knots. The Tornado of the West African offshore area, not

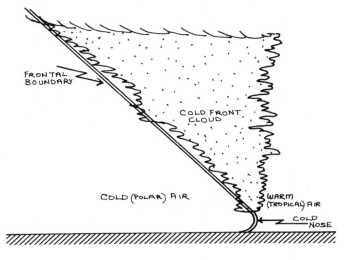

54. The cold nose at a cold front

in any way similar to the overland 'twister' of the southern USA, is also a line squall which moves from east to west during the transition period between the wet and dry seasons in the West African area.

As a general guide it can perhaps be said that the faster the line squall is moving, the stronger the squall will be, but this is not always the case. A 'weather eye constantly lifted' is all one can do in squally conditions and this means frequent all-round checks. At night, listen.

Local Wind Effects

Moving air, the wind, takes the path of least resistance so that it tends to remain over the relatively flat sea for as long as it can before it moves over possibly rough land surface. As a result a wind blowing obliquely onshore will undergo a freshening along the coast (see Diagram 55a). This effect will be more marked if the coastline is formed by cliffs, particularly at a cape, which present a more effective barrier to the wind.

Air tends to move along an estuary or a channel, such as the Solent, rather than obliquely across it: this is known as Channelling (Diagram 55b). When the wind is already blowing along a channel (or estuary) the wind speed over the channel is stronger than that over the land on both sides: this is Funnelling (Diagram 55c). Some estuaries are so sheltered, especially in crosswind conditions, that only light winds prevail in the estuary even though a gale is raging outside. For all these reasons, the wind in an estuary is hardly ever representative of the wind in the open sea.

Sea and Land Breezes

Both sea and land breezes result from the fact that the land warms up more than the sea during the day and cools down more during the night, both being at roughly the same temperature

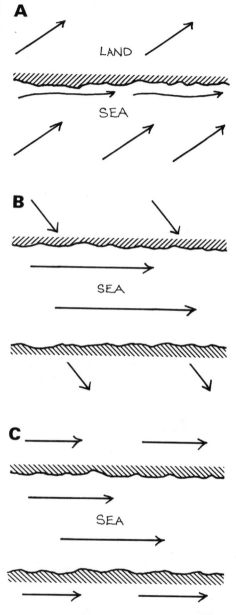

55. Local wind effects: (a) coastal increase in wind speed and change in direction, (b) channelling, (c) funnelling

at about sunset and sunrise. Incoming heat from the sun warms the land surface to a depth of only a few centimetres which results in a large rise temperature in this thin layer. On the other hand, incoming radiation

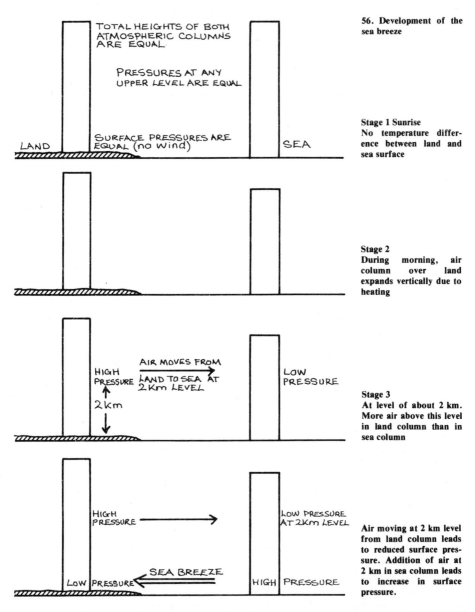

TOTAL HEIGHTS OF BOTH
ATMOSPHERIC COLUMNS
ARE EQUAL.

PRESSURES AT ANY
UPPER LEVEL ARE EQUAL

LAND

SURFACE PRESSURES ARE
EQUAL (no wind)

SEA

56. Development of the
sea breeze

Stage 1 Sunrise
No temperature differ-
ence between land and
sea surface

Stage 2
During morning, air
column over land
expands vertically due to
heating

HIGH
PRESSURE

AIR MOVES FROM
LAND TO SEA AT
2 Km LEVEL

LOW
PRESSURE

2 km

Stage 3
At level of about 2 km.
More air above this level
in land column than in
sea column

HIGH
PRESSURE

LOW PRESSURE
AT 2 Km LEVEL

SEA BREEZE

LOW PRESSURE

HIGH PRESSURE

Air moving at 2 km level
from land column leads
to reduced surface pres-
sure. Addition of air at
2 km in sea column leads
to increase in surface
pressure.

penetrates the oceans to a depth of
about 10 m. The same energy being
shared over a much deeper layer results
in only a small rise of temperature in
the water. By night the process is
reversed. Outgoing radiation from the
ground surface (more particularly
from its top few centimetres) results in
a relatively large drop in temperature
overland. The heat loss over the sea is
shared over a much greater depth and
therefore the drop in its surface
temperature is consequently only
slight.

Strong sunshine, then, leads to large
rises in temperature overland which
results in the air over the land warming
up and expanding. Since the air everywhere over the land is expanding it can
only expand vertically, any sideways

expansion being resisted by the surrounding surface air similarly attempting to expand. It is this vertical expansion which is the triggering effect for sea breezes.

Diagram 56 develops the sea breeze effect in stages, by means of two columns of air, the one representing the state of the atmosphere overland and the other that over the sea. Stage 1 shows the initial equal state at, say, sunrise; temperatures are equal and in this initial state we shall assume that there is no pressure difference, sea to land, i.e. no wind. Due to heating during the morning the column overland expands vertically so that it becomes taller than the column over the sea (Stage 2). The amount of air in each column is the same, however, so surface pressures remain equal. However if we examine the same layer at a given height above the surface then the columns are no longer equal. At, say, 2 km high, there is more air above this level in the column overland than over the sea (Stage 3). There is therefore higher pressure at this level overland. As a result air flows at about the 2 km level from the column overland to the column over the sea in an attempt to neutralize the pressure difference. However the result of this upper level seaward flow of air is that air is added to the sea column and this inevitably leads to higher pressure at the surface over the sea than at the surface over the land (Stage 4). The surface pressure gradient thus established, where there was no such gradient before, leads to a flow of air from sea to land - the sea breeze.

The sea breeze probably affects the region up to 10 miles offshore and sometimes as far as 50 miles inland; its depth is around ½ to 1 km vertically. In tropical regions it sets in at between 1000 and 1100 hours, initially at Force 1 to 2 but soon increasing to Force 3 to 4, and in some areas increases further to 5 to 6 during the afternoon. The direction is at first perpendicular to the coast but the effect of the earth's rotation (the Geostrophic Force) is to gradually veer the sea breeze 40° to 50° in the northern hemisphere (and to back it by a similar amount in the southern) by the afternoon, except within about 10° latitude of the Equator where the Geostrophic Force has no effect. Topography plays a major part in directing the sea breeze. Mountains close to the coast will deflect the sea breeze to run parallel to the coast, and perhaps increase its strength.

The sea breeze is sometimes, especially in its initial stages, very localized. For example, the author has observed yachts off the Isle of Wight shore of the Solent sailing to an onto-the-island sea breeze while simultaneously other yachts on the mainland Hampshire side were sailing to a Solent-to-Hampshire sea breeze on a pure sea breeze day. Later the main north-going sea breeze from the south side of the Isle of Wight dominated over the whole area just as though the Solent, as a sea breeze producing area, had never existed.

There are occasions when in spite of what appear to be favourable conditions, i.e. the land much warmer than the sea (say at least 5°C or 9°F) and no prevailing wind, the sea breeze fails to develop. This may be due to the presence of an inversion which does not allow sufficient depth (possibly 2 km required) to develop the essential vertical circulation. On the other hand, and probably a more fundamental reason for the sea breeze failure, is that if the upper level wind, at about 2 km, is already blowing onshore it will prevent any possible offshore drift at that level and therefore prevent any subsequent onshore movement at the surface.

The sea breeze usually weakens around sunset when the land/sea temperatures become more equal and,

Beaufort wind scale no.	General description	Beaufort's criterion			Appearance of the sea
0	Calm	Calm.			Sea like a mirror.
1	Light air	Just sufficient to give steerage way.			Ripples with the appearance of scales are formed without foam crests.
2	Light breeze	That in which a well-conditioned man-of-war with all sail set and 'clean full' would go in smooth water from:	1 to 2 knots		Small wavelets, still short but more pronounced. Crests have a glassy appearance and do not break.
3	Gentle breeze		3 to 4 knots		Large wavelets. Crests begin to break. Foam of glassy appearance. Perhaps scattered white horses.
4	Moderate breeze		5 to 6 knots		Small waves, becoming longer; fairly frequent white horses.
5	Fresh breeze	That to which she could just carry in chase 'full and by'.		Royals, etc	Moderate waves, taking a more pronounced long form; many white horses are formed. Chance of some spray.
6	Strong breeze			Single-reefed topsails and topgallant sails	Large waves begin to form; the white foam crests are more extensive everywhere. Probably some spray.
7	Near gale			Double-reefed topsails, jib, etc	Sea heaps up and white foam from breaking waves begins to be blown in streaks along the direction of the wind.
8	Gale			Triple-reefed topsails, etc	Moderately high waves of greater length; edges of crests begin to break into the spindrift. The foam is blown in well-marked streaks along the direction of the wind.
9	Strong gale			Close-reefed topsails and courses	High waves. Dense streaks of foam along the direction of the wind. Crests of waves begin to topple, tumble and roll over. Spray may affect visibility.
10	Storm	That which she could scarcely bear with close-reefed main topsail and reefed foresail.			Very high waves with long overhanging crests. The resulting foam in great patches is blown in dense white streaks along the direction of the wind. On the whole the surface takes on a white appearance. The tumbling of the sea becomes heavy and shock-like. Visibility affected.
11	Violent storm	That which would reduce her to storm staysails.			Exceptionally high waves (small and medium-sized ships might be for a time lost to view behind waves). The sea is completely covered with long white patches of foam lying along the direction of the wind. Everywhere the edges of the wave crests are blown into froth. Visibility affected.
12	Hurricane	That which no canvas could withstand.			The air is filled with foam and spray. Sea completely white with driving spray; visibility very seriously affected.

Table 5. Beaufort wind scale and associated average and maximum wave heights.
Wave height data are based on wave recorder statistics. Since maximum wave height depends on
duration as well as wind speed maximum heights for various durations are given. Should a depression

Wind speed range (knots)	Mean wind speed (knots *)	Average height (metres)	Max wave height-any 10 min period	Max wave height any 6 hr period	Max wave height any 48 hr period	Beaufort wind scale no.
<1	0	—	—	—	—	0
1-3	2	—	—	—	—	1
4-6	5	—	—	—	—	2
7-10	9	0.5	0.8	—	—	3
11-16	13	1.1	1.8	2.4	2.7	4
17-21	19	2.0	3.2	4.4	5.0	5
22-27	24	3.1	5.0	6.8	7.5	6
28-33	30	4.5	7.2	9.9	10.9	7
34-40	37	6.7	10.7	14.7	16.3	8
41-47	44	9.3	14.9	20.4	22.6	9
48-55	52	12.3	19.7	27.0	29.9	10
56-63	60	15.5	24.8	34.0	37.7	11
64 and over	?	?	?	?	?	12

move at the same speed as the waves then the 'duration' is considerable. In these circumstances very large waves will arrive *simultaneously* with a very strong wind (see Chapter 18). *Some countries use metres per second in shipping forecasts. To convert m/sec into knots, multiply by 2.

soon afterwards, normally dies away as temperature conditions become reversed ahead of the development of the land breeze. The reader may wish to modify the previous argument, now with a tall column of air over the sea during the evening and a shorter column overland, finally arriving at a land-to-sea movement of surface air overnight. The land breeze is normally weaker than the sea breeze; it sets in during the evening and disappears around sunrise. Its effect may reach 5 miles or so out to sea.

The land sea breeze regime is so regular in some parts of the tropics that fishermen in small outrigged sailing canoes sail out a few miles offshore before daybreak with the last of the land breeze and then, having fished for a few hours, sail home with the sea breeze during the afternoon.

Sea and land breezes make sailing along coasts a possibility when further offshore calms may persist for days. This is particularly so in parts of the Mediterranean in summer.

Katabatic Wind

This is similar in many ways to a land breeze in that it is a night-time effect and is due to cooling of the land surface. It is a hill and valley wind caused by the air in contact with the cooling slopes itself being cooled and, on becoming more dense, sinking to the floor of the valley. If the valley floor itself slopes to the coast then the cold air flows down the slope and eventually out to sea. Katabatic winds are stronger in mountainous areas and their effect off the coast will be more noticeable where the mountains are close to the sea. Under these conditions a katabatic wind may reach Force 5 to 6 close to the coast, gradually decreasing as it fans out to dissipate its momentum by about 5 miles out to sea.

Some remaining well known winds, e.g. Mistral and Bora, are discussed in Chapter 25 on Mediterranean weather.

Beaufort Scale

No discussion on wind, as it affects yachts and larger vessels, can be complete without reference to the Beaufort Wind Scale - a means of estimating the strength of the wind at sea. Originally devised by Admiral Beaufort early in the nineteenth century, it related a scale of numbers from 0 to 12 to the maximum amount of sail that a ship of the line of that time could carry. The scale was later defined in terms of wind speed in knots. Subsequently most textbooks on marine meteorology have published photographs of sea state against the appropriate Beaufort Force so that the state of sea has been used for some considerable time to estimate the wind speed. Despite its shortcomings (discussed in Chapter 18 on waves) the Beaufort Scale is likely to remain in use in this fashion for some time to come even though more and more vessels are being equipped with wind indicating instruments which provide the wind speed (albeit apparent) in knots. Table 5 displays Beaufort Force number against wind speed in knots and state of sea; Beaufort's criteria have been included for interest's sake.

The table also includes probable average and maximum wave heights for most Beaufort numbers; *these are different from the values given in earlier versions of the Beaufort Scale.* The new wave height values are based on wave recorder data (a fairly recent development) which are more realistic and therefore more representative of conditions to be met at sea.

When lying in a sheltered harbour or estuary and planning to sail, an estimate of the wind to be experienced outside the estuary (or later in the day should an early departure be planned) can be obtained by looking at the rate of movement of any low cloud which may be present. By assuming its movement is representative of the geostrophic or gradient wind (that at about 1,000m above the surface) a reasonable estimate of the surface wind outside or later can be made by backing the direction about 20° or 30° and taking about 70 per cent of your assessment of the cloud speed. The skill required can soon be acquired by looking at low cloud movement from the land (particularly the coast) when strong winds are blowing, on as many different occasions as possible. This method is particularly successful if applied to the movement of the first traces of the formation of cumulus cloud caused by morning heating over the land.

One last point on winds, now for the ocean yachtsman: it has often been noted that the trades increase at sunset and that this increase may remain for several hours. No explanation can be offered for this effect, but since it has been noted by experienced yachtsmen it cannot be discounted as an imagined effect due to the onset of night.

CHAPTER 13

Visibility

FOG is another serious hazard faced by yachtsmen. It is probably even more feared than strong winds, especially in busy shipping areas such as the English Channel where larger vessels, equipped with radar, are proceeding at a brisk 'safe' speed. Fortunately sea fog normally occurs with moderate or lighter winds so that a yacht's radar reflector will usually be seen by approaching larger vessels. However, this cannot be relied upon and a good lookout (listening and looking) should always be kept. In any case, sea fog may sometimes occur in stronger winds: then sea clutter (echoes from waves) would probably obscure any radar echo from a yacht.

The Visibility is a measure of the clarity or transparency of the lower atmosphere; it is technically defined as the maximum distance at which a known object can readily be recognized. The clarity of the lower atmosphere is affected by the presence of solid particles (dust, industrial smoke, lifted sand etc) and by water droplets (mist and fog). Serious reductions in visibility can occur with either type of particle, solid or liquid. The most obvious case with solid particles is a sandstorm near arid or semi-arid coasts, though under certain conditions industrial smoke may also cause poor visibility. The 'water' particles appear as mist, fog, drizzle, heavy rain and snow, variously reducing the visibility.

Fog is defined as a visibility of 1 km or less; mist as a visibility between 1 and 2 km. In both cases the reductions are due to water droplets. Haze has no such defined limits, but in practice its upper limit is normally taken as 10 km.

Except in the case of showers of rain or snow, poor visibilities are normally associated with stable atmospheres, i.e. those in which an inversion of temperature exists with warmer air overlying cool air next to the surface. Water droplets or dust and smoke particles are confined beneath the inversion (which may be only a few hundred feet above the surface) and poor visibility may result. In unstable atmospheres (characterized by cumulus type cloud) convection lifts the dust, haze etc high into the troposphere where it disperses.

In general, then, poor visibilities are associated with tropical (stable) air masses, whereas the visibility in polar air masses (unstable) is normally good except in showers.

Haze, Dust, Sandstorms

When an inversion of temperature occurs, industrial smoke is trapped beneath the inversion and visibilities are reduced downwind of the smoke source. Should a substantial inversion exist at a very low level, say below 300 m (see Diagram 57), then poor visibilities of only a few hundred metres may occur downwind of a large industrial region.

Dust, raised by strong winds over arid or semi-arid coasts, may reduce visibilities just offshore to only a few hundred metres. Again, a low-level inversion is required to prevent the dust being dispersed upwards into the atmosphere.

Sandstorms, conversely, require the lower atmosphere to be unstable so that the rising currents due to convection assist the strong winds in keeping the larger sand particles in suspension. Serious reductions in visibility will occur in sandstorms close to some North African, Red Sea and Persian Gulf coasts. Off western Africa, especially when the Harmattan (northeasterly wind) has been blowing strongly over the desert, dust and fine sand is sometimes carried hundreds of miles offshore.

Fog

There are two main types of fog which concern the yachtsman: sea fog and land fog. Land fog, or radiation fog as it is technically known, may sometimes drift off the land to affect inshore areas, estuaries and harbours. It forms when moist air is cooled to its dew-point overnight under conditions of clear skies and light winds, the ideal conditions for maximum cooling of the land surface by radiation. On drifting

seaward, radiation fog normally tends to lift just off the surface and as a consequence there is a significant improvement in visibility within a mile or two of the shore. Land fog rarely spreads more than three miles or so offshore.

Sea fog forms in a somewhat different manner. It results from the cooling of moist air by a colder sea surface. This gives the fundamental condition for the formation, and persistence, of sea fog: that the sea surface temperature must be lower than the dew-point temperature of the air. As a general guide, if the sea surface temperature is 0° to 2°C colder than the dew-point temperature of the air, then sea fog patches result; if more than 2°C colder then sea fog will be widespread.

Another fundamental difference between land and sea fogs is that whereas radiation fog forms when winds are light or calm, sea fog requires a steady breeze to be blowing, normally up to about Beaufort Force 3. A stronger wind normally forms a layer of low stratus cloud just above the surface with moderate visibilities underneath the cloud. However, when the sea surface temperature is much colder than the dew-point temperature (say, more than 5°C, as it often is over the Grand Banks of Newfoundland) dense fog may co-exist with gale force winds!

Due to the requirement that warm moist air moves over a colder sea surface, sea fog is more likely to form in tropical maritime air moving in

57. Reduction in visibility due to smoke trapped under an inversion

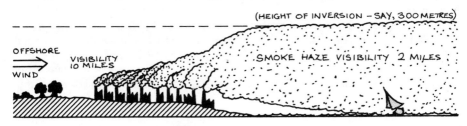

(HEIGHT OF INVERSION – SAY, 300 METRES)

OFFSHORE

VISIBILITY 10 MILES

WIND

SMOKE HAZE VISIBILITY 2 MILES

some poleward direction than it is in any other air mass. The warm sector of a depression is, then, clearly a favoured area for the occurrence of sea fog especially in spring and early summer when, due to the seasonal lag in the recovery of sea surface temperatures, warm moist air may be moving over a much colder sea surface.

Since sea fog is more commonly found in tropical maritime air masses, it follows that the clearance of sea fog requires a change of air mass, i.e. it will normally clear when a cold front passes the area and polar air replaces the tropical air. Sea fog within the warm sector of a depression may sometimes clear by day close to a weather shore (such as the north coast of Brittany) due to the higher temperatures overland. This is usually a temporary effect, and as long as there is no subsequent change of air mass, the sea fog will normally reform, often quite suddenly, soon after dusk.

Global Distribution

Due to the fact that sea surface temperatures within the Tropics are normally higher than the dew-point temperature of the air, sea fog is rare within about 20° latitude of the equator. However, there are two areas where sea fog does sometimes occur: the equatorial region of the eastern Pacific, particularly the vicinity of the Galapagos Islands, and the Gulf of Guinea. In both regions cold Antarctic water is brought northwards by the prevailing currents and as a result the sea surface temperatures there are typically lower than in other regions at the same latitude, sometimes so much so that sea fog develops.

In any region where the prevailing winds are offshore, even obliquely, the effect of these winds is to drive the surface water from the coast. This is replaced by cool (or even cold) sub-surface water, the process of upwelling, which may sometimes be colder

than the dew-point of the air and sea fog results. The process acounts for the occurrence of sea fog along the Californian coast and along the Indian Ocean coasts of Somalia and Saudi Arabia during the Southwest Monsoon.

Sea fog is more common in the temperate latitude westerly belt than it is elsewhere because this is the region where tropical maritime air moves poleward (over cool seas) from the subtropical highs into this low pressure belt. Though sea fog may form in any season, it is more common in spring and early summer when sea surface temperatures are more likely to be below the dew-point temperatures of tropical maritime air. This is due to the fact that the air, originating over the tropical oceans, has already undergone seasonal warming (and acquired higher dew-point temperature) whereas the mid-latitude sea temperatures have only just begun to recover from their minimum values. After all, it is the difference between dew-point and sea surface temperature which is the governing factor for the formation of sea fog and not their individual actual values.

Even within the westerly belt there are certain regions where sea fog is much more common than elsewhere. Those are the Grand Banks area of Newfoundland and a corresponding area in the Northwest Pacific off Japan. In both regions warm poleward-moving currents converge with cold equatorward-moving currents, the two then turning to run side by side in opposite directions (Diagram 75). Tropical maritime air, with exceptionally high dew-points (maintained by its long journey over a warm current), will suddenly be cooled as it eventually flows over the cold current, whose temperature may often be 5°C or more below the dew-point of the tropical air. Widespread fog results, often with fresh or strong winds, and even sometimes with gales.

In polar regions sea fog is more frequent in summer when depressions, which are infrequent in these areas in winter, bring warmer air from lower latitudes into these regions where it is chilled by a colder sea surface. For this reason the vicinity of the sea-ice edge is commonly screened by sea fog in summer.

Another type of fog occurs in the polar regions - Arctic Sea Smoke. This results from very cold air moving over a warmer sea, e.g. in the Arctic a northerly wind with temperatures of about minus 20°C moving off the polar ice over 'warmer' water which will be at least 18°C warmer (since the freezing point of sea water is minus 2°C). Evaporation from the sea surface soon super-saturates the cold dry air and the excess is condensed out as fog which has the appearance of steam rising from the surface in the vigorous convective currents. Arctic sea smoke may sometimes reduce visibility to only a few metres, and though it is commonly only two or three metres in depth it sometimes extends as high as twenty metres, thus obscuring the tops of masts of other vessels in the vicinity.

The estimation of visibility at sea is extremely difficult except when it is good and a sharp horizon can be seen, or when it is very poor and then can be assessed in boat lengths. In general, it appears to be a natural (and not unseamanlike) characteristic to under-estimate the visibility when it lies between 1 and 5 miles or thereabouts. Again, the visibility appears to be worse in the direction towards the sun (up-sun) than it does in the opposite direction; conversely, it appears better up-moon. At night, the presence of sea fog can be noted from the 'loom' around the vessel's lights.

The prediction of sea fog by a yachtsman at sea is almost impossible without instruments to measure the difference between dew-point temperature and sea surface temperature. The measurement of the latter is relatively easy, but very few yachts, if any, will be equipped with a psychrometer (see Chapter 16) which measures humidity and dew-point temperature. In areas outside of the coverage of radio forecasts the seafarer will have to rely on the previous discussion, together with local information from sailing directions etc., and his own assessment. In areas where shipping forecasts are available or there are local meteorological offices (on airfields etc), the professional advice should be heeded. Sea fog may develop very suddenly; it has often embarrassed small boats within only a mile or so of the coast, many of which have not even been equipped with a compass!

The method of forecasting sea fog, and in fact the workings of the meteorological forecasting service in general, will be described in the next chapter.

CHAPTER 14

Preparation of Forecasts

ANY form of scientific prediction consists of two stages, an analysis stage followed by a predicting, or prognostic, stage: the accuracy of the latter depending to a large extent on the completeness of the former.

In meteorology the analysis stage consists of determining the state of the atmosphere at a given time, i.e. where the anticyclones and depressions are located, where there are areas of good and bad weather, where the air mass boundaries (fronts) are, where there are strong winds, and where the atmosphere is stable and where unstable. To arrive at this stage, a vast amount of basic information is required over a large area. The size of this area depends on the length of the forecast period. If this is only an hour or two, then an area of a few hundred square miles is sufficient; if for twenty-four hours, then a major portion of the appropriate hemisphere; and if for several days then practically the whole hemisphere is involved.

This is because weather events, except for short-period showers or thunderstorms, rarely happen in isolation. Due to the general circulation of the atmosphere, a major change of weather over one large area (say the eastern states of the USA) normally produces a chain reaction which often results indirectly in changes of weather over areas thousands of miles further east, such as western, or even central, Europe, within a day or two. These resultant changes are not necessarily in the same sense; a deterioration over the eastern USA may result in an improvement in the weather conditions over western Europe.

Most shipping forecasts are for twenty-four hour periods so we shall concern ourselves with the methods by which these forecasts are professionally produced. The appendices supplement this discussion.

The basic data on which the analysis for such forecasts depends, consists of thousands of surface observations carried out by experienced observers, together with hundreds of upper-air soundings of the atmosphere obtained from radio-sondes. (These are balloon-borne instruments, transmitting temperature, humidity and pressure data throughout the atmosphere up to heights of about 30 km - see Appendix 4.) These observations, both surface and upper air, are made at least twice per day over the whole globe; the time standard is GMT and the main observation times are 0001 and 1200 GMT. Somewhat less data is available at 0600 and 1800 GMT (especially upper-air data). The data at these four times normally form the basis for six hourly shipping forecasts, and we shall look at the data for 1200 GMT.

At this hour many thousands of observers all over the world will have made surface weather observations and perhaps half an hour or so earlier many hundreds of radio-sonde balloons will have been released and will now be transmitting information on the state of the upper air. By means of a high-speed global communications network nearly all of this information, coded into internationally adopted forms,

will be available at many meteorological offices in practically every country within an hour or two, and much of it in some national headquarters within a matter of minutes!

Surface observations are plotted onto surface charts in a manner used by practically every country while upper-air data from radio-sondes is plotted onto upper-air charts at various levels up to about 15 km the whole forming a three-dimensional array of the basic data.

In analysing surface charts, isobars

58. Analysed surface chart for 0001 GMT 9 Aug 1979. By taking $\frac{2}{3}$ geostrophic wind speed in warm sector ($\frac{2}{3} \times 33 = 22$ kt – see Chapter 11) and moving the low in the same direction as the warm sector isobars, this gives a movement of 264 n. miles towards ESE in 12 hr. The forecast position is almost exactly the actual position (marked X) at 1200 GMT on the same day.

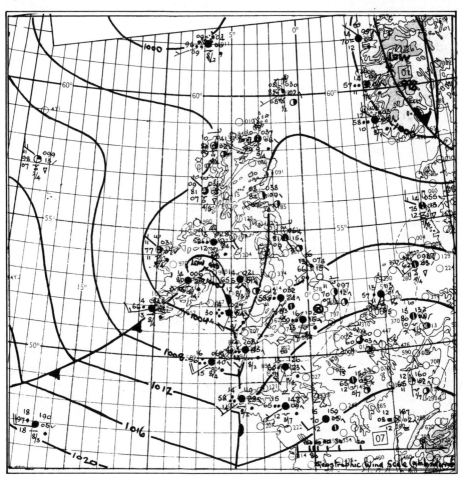

(lines of equal pressure) are drawn and these automatically reveal the positions of depressions and anticyclones; fronts are drawn in at the boundary between the tropical and polar air masses. Where data are sparse such as over the oceans and deserts, satellite pictures (Appendix 5) are used to locate fronts and depressions. The pattern of highs, lows and fronts displayed on analysed surface charts is commonly referred to as the Synoptic Situation or the General Synopsis. A small section of an analysed chart, showing plotted observations, isobars and fronts, is shown in Diagram 58.

Lines, closely approximating to isobars, are drawn on upper-air charts at various levels up to about 15 km. The pattern of these lines reveals the position of upper highs, lows, troughs and ridges. These, especially in the levels between about 5 and 12 km, are important for they not only tend to govern the movement of surface anticyclones and depressions but also their development and decay.

The movement and development of depressions and anticyclones forms the basis of meteorological prediction. Weather, in the form of a discrete rain shower at one end of the spectrum, or the vast area of rain associated with a depression at the other, does not last forever. The shower has a lifespan of up to an hour or two, while the latter a few days or perhaps a week or so. In predicting the movement and development of surface highs and lows by means of the patterns analysed on upper-air charts, allowance must also be made for the fact that these upper-air patterns are themselves altered by the changing surface patterns beneath them.

The prognostic chart showing the positions of depressions, fronts and anticyclones forms the basis of the shipping forecast. The winds and weather in each area will be forecast according to the movement and development of highs and lows across the area. The forecast visibility will depend on the air mass(es) and their boundaries (fronts) affecting the area, with special consideration being given to the comparison between the dew-point temperature of the air mass and the sea surface temperature (sea fog). The visibility will also depend on the stability of the atmosphere since stable air masses often form low-level inversions of temperature with their attendant haze, and unstable air masses usually contain showers with associated moderate or even poor visibilities.

This much-simplified account of the production of forecasts even so reveals something of the complex nature of the problem. Though these methods still persist in many meteorological services, in the more advanced nations forecasts are produced by numerical means using a totally different approach which can only be handled by a large and powerful computer. The basic data, thousands of surface observations and hundreds of upper-air observations, is analysed directly by the computer in terms of interpolated values of pressure, north and east components of the wind, humidity, temperature and various other relevant factors for a number of evenly spaced grid points over an area of many thousands of square miles, at ten or so levels between the surface and about 20 km. The method normally allows for human intervention which includes adding late data and estimating values where there is little or no data. (These estimates will be based on recent data, satellite pictures and aircraft reports.) Several complex equations, collectively describing the behaviour of the atmosphere with the passage of time, are then solved for each grid point at each level for short time intervals, typically 15 minutes. This is repeated over and over again until finally new values of pressure, temperature, humidity and wind are obtained for each grid point at intervals, most importantly 24 hours ahead. Isobars, drawn from the new forecast pressure values, show the forecast position of depressions and anticyclones; fronts are located along bands of high humidity. Once again the indicated movement and development of the depressions, anticyclones and fronts from the basis of the shipping forecast.

CHAPTER 15

Shipping Forecasts

THOUGH there are some national differences in the form of transmitted shipping forecasts, most begin with gale warnings and the synoptic situation (the positions of highs, lows and fronts and their forecast movement and development) followed by forecasts of wind, weather and visibility for each sub-division of the total area covered by the shipping forecast. Since forecast times and frequencies are liable to be altered from time to time there is no point in listing them here. They are in the Admiralty List of Radio Signals, Vol. 3 and 3a, List of Radio Services for Small Craft NP280, similar publications from other maritime nations, or nautical almanacs.

It is often the way of things on a yacht at sea, that some manoeuvre, be it sail-changing or the avoidance of a collision situation, has to be undertaken at shipping forecast transmission times. An alarm clock is therefore useful to announce that the forecast, or any other scheduled transmission, is about due. The system becomes much more fail-safe if a cassette tape recorder is also used for then the contents of a transmission may be repeated at leisure after the excitement of a particular manoeuvre has gone by. A tape recorder will also greatly facilitate the subsequent writing down and plotting of the shipping forecast.

The following description relates to the content of the shipping forecast issued by the UK meteorological headquarters at Bracknell via the BBC and various Coastal Radio Stations, and this can perhaps be taken as representative of the style of most national shipping forecasts.

The forecast begins with a summary of gale warnings already in operation. (New gale warnings are normally transmitted at the earliest opportunity after receipt.) This is followed by the general synopsis which describes the positions of depressions, fronts and anticyclones at a given time and their forecast positions 24 hours later. Next come forecasts of wind, weather and visibility in each sea area around the British Isles followed by the latest weather reports from a number of coastal stations. It is as well, here, to differentiate between reports and forecasts. The term 'report' describes an actual observation of the weather at a given time - a factual statement - whereas a forecast is a prediction of expected weather; the two terms should not be confused, though they often are.

Normally, transmissions are restricted to five minutes or less so there is obviously a need for the originator to economize on words and to avoid ambiguity or words with vague definitions. Some terms which in common usage have very loose definitions are given much more precise definitions in shipping forecasts, particularly with regard to timing. For gale warnings, the terms relating to timing are:

Imminent	means within 6 hours of time of issue
Soon	means in 6 to 12 hours of time of issue
Later	means more than 12 hours beyond time of issue.

The terms used for visibility are:

Fog	means less than 1100 yards (1000m)
Poor	means 1100 yards to 2 nautical miles
Moderate	means 2 to 5 nautical miles
Good	means more than 5 nautical miles.

The following terms indicate the speed of pressure systems, fronts etc:

Slowly	means less than 15 knots
Steadily	means 15 to 25 knots
Rather quickly	means 25 to 35 knots
Rapidly	means 35 to 45 knots
Very rapidly	means more than 45 knots.

The writing down of the shipping forecast will be made much easier if specially prepared forms are used. A number now available, published by the Meteorological Office, the Royal Meteorological Society and the RYA, display the forecast areas and coastal stations in map and tabulated form which makes for rapid entry of the forecast. Nevertheless, since the forecast is read out at normal speed and not dictation speed some form of 'shorthand' will have to be used.

Wind direction, weather and visibility can be reduced to initial letters, e.g. W for westerly, R for rain and M for moderate visibility; since these elements are always broadcast in that order there should be no ambiguity. The passage of time may be indicated by an arrow, and wind direction changes by a counter-clockwise or clockwise arrow. For example 'Plymouth. Westerly Force 4 to 5 backing southerly Force 6 later, rain later, good becoming moderate later' can be reduced to:

$$W4\text{-}5 \; \underset{\curvearrowleft}{} S6, \; \rightarrow R, G \; \rightarrow M$$

When a depression is expected to pass over a given area the forecast wind direction may be given as 'cyclonic'. This, and its associated wind force, may be entered as ⑦

In most cases, shipping forecasts will be substantially correct, but due to the complex problems associated with predicting the behaviour of the atmosphere, in particular the movement and development of transient eddies (depressions), some forecasts will be incorrect. Normally the inaccuracies will have been detected by the time of the next forecast six hours later. The prudent course is to take down each forecast and also to monitor the weather changes in the six-hour interval between each forecast by 'lifting a weather eye'. (Here the guidance of Chapters 8 and 9 may be found useful.) This monitoring, of course, reveals changes in one's immediate vicinity. A more general check on the forecast may be made by plotting a weather chart using later information than that on which the shipping forecast is based, i.e. by using the coastal station reports which are normally made three to four hours later than the basic data. This is sufficient time to detect unexpected developments.

It is well worth plotting such charts since a series will show the development of the weather in pictorial fashion (and in any case, the ability to do so is often a requirement for yachtsmen's and other examinations). But how should this be done?

Firstly, the coastal station reports have to be taken down. These reports contain, in order, wind direction and Beaufort Force, weather (if any 'significant weather', i.e. rain, fog, drizzle, snow etc), visibility (in nautical miles but in yards below 2 miles), the atmospheric pressure in millibars, and the pressure tendency, i.e. the character of the pressure change over the past three hours such as falling slowly, rising quickly, steady, etc. Similar abbreviations to those in the shipping forecasts may be used for coastal station reports, but additional abbreviations will have to be included particularly for 'pressure tendency'. This is perhaps most easily recorded as a symbol representing a graph of pressure against time as shown in the following examples.

'Royal Sovereign, southeast 2, fog, 1000 yards, 1022 millibars, falling slowly may be shortened to

SE2   F   1000   1022 $\diagdown$

'Bell Rock, Northeast 6, heavy rain, 3 miles, 997, falling more slowly' may be shortened to

NE6   HR   3   997 $\diagdown$

(The term 'more slowly' means that the rate of fall (or rise) is now less than it was; 'more quickly' means that the rate is greater than it was.)

Again, the terms used for pressure tendency have precise meanings, as follows:

'Steady' means a change of less than 0.1 mb in the past 3 hours.

'Rising or falling slowly' means a change of 0.1 to 1.5 mb in past 3 hours. 'Rising or falling' means a change of 1.6 to 3.5 mb in past 3 hours. 'Rising or falling quickly' means a change of 3.6 to 6.0 mb in past 3 hours. 'Rising or falling very rapidly' means a change of more than 6.0 mb in past 3 hours.

Clearly the last definition is significant, especially with a falling barometer, but sometimes too when rising, since this will normally mean a blow. Though it is difficult to lay down rules relating pressure tendencies with subsequent wind strengths the following guidelines may be useful, especially with a falling barometer.

1. When the pressure tendency reaches 6 mb (in 3 hours) there is a high probability that a gale, Force 8, will follow in a matter of hours.

2. When the pressure tendency reaches 9 mb there is a high probability that a severe gale, Force 9, or a storm, Force 10, will follow.

Having taken down the coastal station reports, the next step is to plot them onto the chart, boldly and in ink. More symbols will have to be used in meteorological practice.

Wind direction, which is that *from* which the wind is blowing, to be plotted as an arrow towards the station; each Beaufort force as half a feather on the 'clockwise' end of the shaft, e.g.

NW   3   as

and SW by W   6 as

Weather should be plotted on the left-hand side of the station (or as near as the wind plot will allow) using the shorthand symbols shown in Table 6.

A further distinction between intermittent (began less than 1 hour ago) and continuous precipitation (began over an hour ago) is used in meteorological practice but for the purposes of this discussion there is little benefit from using this distinction. Note that intermittent precipitation and showers are different, however, because the former falls from layer cloud (nimbostratus) while the latter falls from convective cloud (cumulonimbus) although their duration may be similar.

Visibility should be plotted on the left-hand side of the station outside the weather symbol, if necessary noting Y for yards and M for mile, though there should be no ambiguity since low figures in yards will have a fog symbol alongside (see examples later). Pressure is plotted top-right and pressure tendency on the right side of the station.

The following examples are offered to amplify the previous discussion, firstly as transmitted, then as written down using abbreviations, then as plotted. With only a little practice the second stage can be eliminated and the coastal station report plotted straight from the radio or tape recorder on to the chart. The yacht's own weather for the same time as the coastal station reports should also be plotted.

Ronaldsway Southwest by west 3, fog, 50 yards, 1007 mb, falling.

SW'W 3 F 50 1007 \

Channel Light Vessel Southeast by east 8, moderate rain 3 miles, 993 mb, falling.

SE'E 8 MR 3 993 \

Bell Rock East 6, rain showers, 8 miles, 1013 mb, steady.

E 6 R Sh 8 1013 —

Valentia Southwest 5, slight drizzle, 2 miles, 1010 mb, falling slowly.

SW5 SD 2 1010mb ﹀

2﹜ 1010

To make the plot more complete, the 'at first' forecast winds should be plotted in the appropriate sea areas. Now plot any information from the general synopsis, i.e. positions of highs, lows and fronts, making due allowance for the time interval between that of the synopsis and that of the coastal station reports (usually three hours). Since the actual position of, say, a depression is given and also its forecast position 24 hours later, its position at the time of the coastal station reports (the time of our weather

	Slight	Moderate	Heavy	Showers
Rain	•	• •	• • •	ᵥ̇
Drizzle	﹐	﹐ ﹐	﹐ ﹐	
Snow	✳	✳ ✳	✳✳ ✳	✳ ᵥ

Hail	**Thunderstorm**	**Fog**	**Distant Fog**	**Mist**	**Haze**
⌃̱ᵥ	℟	≡	(≡)	=	∞

Table 6. Symbols for plotting on weather charts

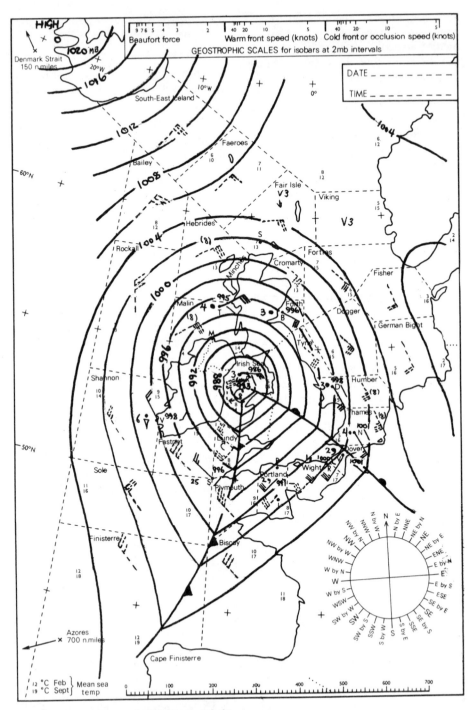

59. Example of yachtsman's weather chart. (Acknowledgement is due to the Royal Meteorological Society and the RYA for the use of the background chart.)

chart) can be obtained by taking 3/24 of the expected track. This information should be entered in pencil since it is possibly going to be adjusted after the coastal station reports.

Next sketch in the isobars (lines joining places of equal pressure), remembering that the wind directions are very roughly parallel to the isobars and that (in the northern hemisphere) the low pressure is on the left when running with the wind. (Sometimes the wind at the coastal station may be almost perpendicular to the isobars. This is due to the topography in the vicinity or to sea/land breezes and such effects.) When sketching the first isobar select a value which is common to at least two coastal stations and then sketch in the remainder, at 2mb intervals, according to the pressures at the other coastal stations, their winds and winds over the sea areas. (By convention, isobars are normally drawn for even values e.g. 1000, 1002, 1004 mb etc.) Some blank charts, such as that prepared by the Royal Met.Soc./RYA, include a Geostrophic Scale for Beaufort Wind Force, and the speed of warm and cold fronts. This scale may be used to give the distance between isobars when drawing isobars over the sea areas by using the 'at first' forecast Beaufort Force. The method is to place one point of a pair of dividers on the extreme left of the scale marked 'Beaufort Force' and the other on the forecast Beaufort number; the dividers are then placed over the sea area, at right angles to the forecast direction, to give the correct isobar spacing. The positions of the fronts should be adjusted according to the coastal station reports by using Table 3 which lists the weather changes across fronts. When satisfied with the positions of fronts and isobars, these can then be drawn in more boldly.

The analysed chart in Diagram 59 is based on the following relevant portions of a forecast.

General Synopsis at 0700 BST: Low, South Irish Sea, 983, moving steadily northeast to Viking, 986, by 0700 tomorrow. Associated frontal trough centre to West Sole moving rather quickly east, to clear all areas by this time tomorrow.
High West Iceland, 1020, almost stationary.

The area forecasts for the next 24 hours:
Viking.   Variable 3 ...
Forties.   Southeast 5 or 6, but variable 3 in east at first ...
Cromarty, Forth, Tyne, Dogger. Southeast 6 or 7 ...
Fisher, German Bight.   South or Southeast 3 or 4 ...
Humber,   Thames.   South   or Southeast 6 or 7, locally gale 8 ...
Dover, Wight, Portland. Southwest 6 or 7 ...
Plymouth.   Northwest 6 or 7 ...
Biscay.   Northwest 5 or 6, but Southwest 4 in south at first ...
Finisterre.   Northwest 4 or 5 ...
Sole.   Northwest 5 or 6 ...
Lundy, Fastnet. Northwest 6 or 7
Irish Sea.   Cyclonic   becoming North 7 or gale 8 ...
Shannon.   Northwest 4 or 5 ...
Rockall.   North 5 or 6 ...
Malin,   Hebrides.   North   or Northeast 6 or 7, locally gale 8 ...
Bailey.   Northeast 4 or 5 ...
Fair Isles.   East 5 but variable 3 in north at first ...
Faeroes.   Northeast 3 ...
Southeast Iceland Northwest 4 or 5...

(In all these sea areas the 'later' changes in winds, except for changes early in the period, together with weather and visibility, have been omitted.)
Weather reports from coastal stations for 1000 BST:
Tiree. East-northeast 6, 4 miles, rain, 995, falling.
Bell Rock. East-southeast 6, 3 miles, 996, falling quickly.

Dowsing. South-southeast 6, 3 miles, rain, 998, falling quickly.

Noordhinder. South-southeast 5, 4 miles, rain, 1001, falling quickly.

Varne. Southwest 7, 2 miles, drizzle, 1001, falling more slowly.

Royal Sovereign. Southwest 7, 1 mile, drizzle, 1000, falling more slowly.

Channel Light Vessel. Southwest by west 6, 2 miles, drizzle, 997, steady.

Scilly. Northwest by west 5, 25 miles, 996, rising.

Valentia. Northwest by north 5, 6 miles, rain showers, 998 rising.

Ronaldsway. Easterly 7, 3 miles, rain, 986, falling quickly.

In constructing the met chart, then, the coastal stations are plotted boldly in ink and the 'at first' forecast winds for each sea area. Then plot positions of the depression, anticyclone and frontal trough, adjusted for their movement during the time interval between the general synopsis and the time of the coastal station reports. Now begin sketching in the isobars, by choosing a value which is fairly easy to fit to the observations plotted. In this case 996 mb is as good as any and having sketched this in the remainder may be sketched in according to the station reports and the 'at first' forecast winds for each sea area. Adjust the position of the centre of the low according to this sketched analysis and if necessary adjust the position of the 'frontal trough' according to weather, wind veer, improvement in visibility, pressure change - it is of course a cold front.

Though a warm front was not mentioned in the general synopsis (broadcast time available would not allow) a warm front can be drawn in separating the 'North Sea south-easterlies' from the 'Channel southwesterlies'. (Note, too, that the visibilities deteriorate further in the warm sector, that the rain gives way to drizzle, and that the pressure falls are less steep or even steady off. Table 3 will be of assistance here.) Remember that the isobars should be drawn as straight as possible in the warm sector which automatically means that they must be Vee'd out or troughed out (away from the low pressure centre) at each front - clearly shown in Diagram 59.

Once satisfied with the positions of isobars, fronts and centres of low and high pressure, all of these features may now be drawn in more boldly. Even the professionals will erase several sketchy attempts at their analysis before finally drawing in the completed version.

A comparison between the positions shown (chiefly of the low pressure area) with the estimated forecast position will immediately test the initial accuracy of the forecast. In our example, according to the forecast the low ought to be in the Central Irish Sea at 1000 BST and so it is. Also the central pressure of about 983 mb is in line with forecast developments. Therefore the whole forecast for the 24 hour period will probably be accurate. If for example our 1000 BST chart had revealed that the low had moved into the Bristol Channel then the track predicted is already seriously in error and so the forecast will be incorrect especially in the Channel and southern North Sea where the winds will be much stronger and the change from southeasterly to southwesterly will be delayed. Both effects would be increased if the central pressure had dropped to, say, 976 mb by this time. On the other hand, had our 1000 BST analysis indicated that the low had accelerated northwards and was already in sea area Malin, then winds over southern and eastern sea areas would be less than forecast, whereas those over northern and western areas would be much stronger.

A series of six-hourly charts will show the development of the weather over the general area, i.e. the movement of lows and whether they are deepening or filling and of highs and whether they are weakening or intensifying.

It has already been said that most forecasts are correct: some are not. Some of the inaccurate ones, fortunately a rare event, are dramatically incorrect. In these circumstances a yachtsman may find himself off a potentially dangerous lee shore. It is then no use blaming the forecast; he will have to get the vessel out of the potentially hazardous situation. By listening to the forecast, plotting a met map and keeping a weather eye lifted, a yachtsman will be able to detect that the forecast developments are going wrong and will be much more able to make correct decisions regarding the safety of his yacht.

Only a little practice will reduce the time required to produce a chart to a very few minutes. This practice does not have to occur at sea; it can be carried out at any time. Having plotted a map from, say, the early morning shipping forecast, watch the development of the weather across your area during the day; if a front is expected to affect your area note the sky changes as it approaches and the changes as it passes by. The general synopsis for the forecast given at the same time next day will itself be a check on the accuracy of the original forecast.

Apart from the shipping forecasts, several other forecasts, mainly for land areas, are broadcast from time to time but, better still, a last-minute telephone call to the nearest meteorological forecast office when the weather developments over your area of interest can be discussed with the duty forecaster will give the latest, and therefore most accurate, information.

For yachtsmen on oceanic passages, less detailed forecasts are available for most regions, except parts of the Southern Ocean, south of about 50°S.

These forecasts give the synoptic situation and its development over the next 24 hours over large ocean areas; some will also give forecasts of the wind over these areas. They are well worth listening to especially in the westerly wind belts, for a series of these will give depression tracks over a day or two so that, by adjusting the yacht's course, following winds may be used with good distances run instead of lying-to for long periods until the headwind and seas give way to more favourable conditions. The central areas of stationary highs with their very light and variable winds may also be avoided in this way.

Some countries, as part of their oceanic weather bulletin, include a detailed analysis, in code, giving the positions of depressions and anticyclones and their central pressures, the positions of fronts and the positions of sufficient isobars to complete the chart. This analysis is broadcast in Morse by W/T but as the code is in figures the task of taking this down is not as daunting as it would first appear, especially if a tape recorder is used. Again, an hour or so of effort here, each day, may save several days on total crossing time. The code used is given in ALRS Vol 3.

Within the tropics, where the main concern is tropical revolving storms, there are numerous main centres such as Guam, Manila, Auckland, Darwin, Mauritius, Delhi and Miami which are responsible for issuing warnings of tropical revolving storms. Since these warnings are based mainly on satellite surveillance, it is very unlikely that a storm will escape detection.

Full details of oceanic forecasts, coded analyses, stations and frequencies are given in the Admiralty List of Radio Signals Vol. 3 and 3a, and in similar publications from other maritime nations.

CHAPTER 16

Meteorological Instruments

THE most common scientific instruments found aboard yachts are wind and pressure measuring instruments - anemometers and barometers. Several types of each instrument are available with wide variations in cost. The most expensive wind indicators are those fitted at the masthead - a vane for direction and rotating cups for speed - with dials which can be seen by the helmsman and possibly repeated in the navigation area. These dials indicate wind direction relative to the boat's head and the apparent wind speed. Towards the other end of the cost sale is a hand-held cup anemometer which is held to windward and registers apparent wind speed on a dial. Cheaper still is a Ventimeter in which a small float responds to pressure changes caused by variations in wind speed. The wind speed is indicated by the position of the float against a graduated scale.

Both these hand-held instruments record wind speed only; they do not register direction. The true wind direction may be estimated by noting the line of the wave crests and taking a perpendicular to it since waves run downwind with their crests at right angles to the wind direction. It is important here to observe only the wind waves and not swell (see Chapter 18).

A barometer should be a standard part of a yacht's equipment. Though there are several types in use, the most suitable for a yacht is a standard aneroid barometer, such as those seen hanging in the entrance hall in homes. The aneroid barometer consists of a corrugated hollow sealed disc in which there is a partial vacuum. The disc expands and contracts when the pressure falls or rises, respectively, and the resultant slight movements are magnified by a system of rods and pivots terminating in a pointer which indicates pressure on a circular scale. Their chief use is not so much to give the actual pressure, but to indicate the pressure changes between hourly or even more frequent readings. The actual pressure values are also important, however, especially when plotting the yacht's observation when making a weather map. To ensure the accuracy of the barometer, it should be calibrated against that of the nearest official meteorological office. A simple request for the sea-level pressure *now* at your location will suffice; the yacht's barometer can then be re-set, if necessary, to the value obtained. This should be done at not less than monthly intervals. Before reading the barometer tap the glass face gently with the finger tips to reduce the effect of friction within the various linkages inside the instrument. When keeping a log of pressure readings there is no need to bother with the 'dead' hand which is provided for comparative purposes only.

A thermometer, preferably mercury-in-glass, is another useful instrument on a yacht, not only for measuring air temperatures both on and below decks but also for measuring sea surface temperatures. When measuring the outside air temperature the thermometer must be kept in the shade as strong sunlight will cause it to overread. To measure the sea surface temperature, collect some sea water in a bucket and then keep the bucket in the shade while the thermometer settles down to the sea temperature.

Unfortunately, there is no easy method to measure the humidity, or more particularly, the dew-point temperature of the air. On larger vessels and at land-based observing stations a wet-bulb thermometer (a standard thermometer whose bulb is covered by a muslin sock from which a wick runs into a supply of distilled water) is kept together with a dry-bulb thermometer in a ventilated box or screen. Water is evaporated from the muslin of the wet-bulb thermometer, the latent heat required for this evaporation coming from the thermometer bulb which therefore cools. The amount of evaporation depends on the initial humidity of the air flowing through the screen. If this is already high then only a little evaporation takes place and the wet-bulb temperature will be only a little below the dry-bulb temperature. If the air is dry then evaporation will be considerable and the resultant wet-bulb temperature will be well below the dry-bulb temperature. By entering the values of dry and wet-bulb temperatures in a special humidity table, the dew-point temperature is obtained.

A smaller version of this system of measuring the humidity of the air is available as a hand-held whirling psychrometer. A pair of thermometers (one dry- and the other wet-bulb) are housed in a frame almost identical to a football rattle. After moistening the wet-bulb, the instrument is whirled around for a minute or so and then the dry- and wet-bulb temperatures are read; the values are then entered into a table to give the dew-point temperature. The latter when compared with the sea surface temperature is an excellent sea fog predictor. When the dew-point temperature is less than the sea surface temperature, sea fog is unlikely. When it is up to 2°C higher than the sea temperature patchy sea fog may be expected, and when it is more than 2°C higher then sea fog will be widespread (see Chapter 13).

CHAPTER 17

Tropical Revolving Storms — Hurricanes, Typhoons and Cyclones

IN planning ocean passages into tropical regions, most yachtsmen carefully select those months when tropical revolving storms are either unknown or are at a minimum frequency, even though this may sometimes mean delaying departure for several months. There is good reason for this since tropical revolving storms are the most destructive weather hazard to be met at sea. Of those yachts which have been unfortunate enough to encounter a mature storm only a very few have, miraculously, survived. It should be reckoned that yachts will founder in such storms and therefore they must be avoided.

Tropical Revolving Storm is the name given to an often violent, cyclonic circulation of air which occurs within the tropics. A tropical revolving storm is similar in some ways to a depression of higher latitudes. The main difference is that its area is normally much less and, since the central pressures may be about the same as a depression, the pressure gradients are much more steep and therefore the winds are much stronger. Tropical revolving storms are classified according to the maximum wind speed near to the centres:

While speeds remain at Force 7 or below, it is known as a Tropical Depression

With maximum speeds between Force 8 and Force 9, as a Moderate Tropical Storm.

With maximum speeds between Forces 10 and 11, as a Severe Tropical Storm with maximum speeds Force 12 and above, as a Hurricane (or other local alternative name).

This is an almost internationally adopted classification, but in some areas different terms are used, e.g. in the Indian Ocean sector the second and third stages are known as Moderate and Severe Cyclonic Storms. The final, mature stage is known as a Hurricane in the Western Atlantic and Eastern Pacific (on both sides of Central America) and in the Southwest Pacific, as a Typhoon in the western North Pacific, and as a Cyclone in the Indian Ocean sector.

In the following description, the mature stage will be referred to as a Hurricane except in the accounts by regions where the local name will be used.

Formation
Tropical revolving storms require the following conditions for their formation:

—a large warm oceanic area
—a trough-like disturbance in the upper air pattern

—there should be no very strong winds in the upper air (jet streams) over the area

—the latitude must be more than 5° from the equator.

Their formation is normally limited to those parts of the tropical oceans where the sea surface temperature is at least 27°C (80°F). The surface air here has probably been over the ocean for some time so it is not only warm, it is also moist - the ideal surface conditions for strong convection. However, since their formation is localized, some other triggering action must be present. This is supplied by trough-like disturbances called 'easterly waves' which occur from time to time in the deep easterly flow which normally predominates from the surface trades upwards throughout the depth of the troposphere. The convergence of warm, moist air into these easterly waves results in vigorous convection and the production of large cumulus and cumulonimbus over the region around the easterly wave.

The easterly waves, for reasons which are not fully understood and in any case need not concern us here, do not form when strong winds (jet streams) exist in the upper troposphere, over the area under consideration.

So, within our easterly wave, vigorous convection is occurring which, due to the condensation of moist air into cumulus and cumulonimbus clouds, results in the release of vast quantities of latent heat. This is converted to kinetic energy which is harnessed by the earth's deflecting force - the Geostrophic Force - to form a cyclonic rotation. It is here that the relevance of the fourth condition should be explained. The Geostrophic Force (or at least its horizontal component) is zero at the equator and negligible within about 5° of it. This is why tropical revolving storms cannot form at latitudes of less

than 5° from the equator.

Continued, and perhaps even more vigorous, convection as the easterly wave moves westwards under the influence of easterly winds will make the cyclonic rotation more and more violent.

Movement and Development

Most tropical storms are born over the eastern sides of the tropical oceans, and normally move slowly westwards at about 10 knots under the influence of the easterly winds which prevail throughout the whole depth of the troposphere in these latitudes. This westward movement occurs on the equatorward side of the sub-tropical high pressure belt. In summer and autumn (when hurricanes normally occur) the sub-tropical high pressure areas are confined to the oceans so that as the storms reach the western sides of the oceans they tend to be steered poleward around the western side of the high pressure region before moving away northeastwards (southeastwards in the southern hemisphere) and then usually accelerating to about 20 knots. The path thus described is typically roughly parabolic; the 'elbow' in the path is known as the Point of Recurvature. A few representative hurricane tracks are shown, by areas, in Diagrams 60-65. These not only indicate the parabolic path sometimes followed, but also show the high degree of variability in the track that an individual storm may follow. It is this variability which is one of the main characteristics of tropical revolving storms. They must always be treated with caution for even those which seem to be behaving in the 'typical' fashion may suddenly and dramatically turn through a right angle and charge away polewards. Others may stop for a day or two before resuming the original track while others perform a small loop or temporarily back-track before moving on.

Tropical revolving storms usually develop from 'disturbances' which are often no more than cloud clusters perhaps one or two hundred miles across (Plate 15). As they move westwards in the latitudes under discussion, say 5° to 15°N and S, some may take on a definite cyclonic circulation - the Tropical Depression stage - and some of these may acquire more vigorous circulations - the Tropical Storm or Severe Tropical Storm stage. A few of these will develop into mature hurricanes. The transition from cloud cluster to mature hurricane stage may occur within 24 hours.

A satellite picture of a mature hurricane is shown in Plate 16. The main circular cloud mass in a hurricane may attain a diameter of about 500 miles and into this several spiral feeder bands converge. At the centre of the cloud mass there is an almost cloud-free zone; this is the Eye of the hurricane and may be as much as 30 miles or so in diameter. (Some satellite pictures have revealed a complex temporary double-eye structure, probably one vortex centre taking over from another within the small central section of the storm.) The clouds within the hurricane are thicker as the centre is approached. Rain falling from the clouds becomes heavier until it is torrential within an annulus of very thick cloud which itself contains the cloud-free eye of the hurricane; the inner edge of this annular cloud mass is called the Eye Wall.

Under the areas swept by the storm's feeder bands winds of at least Beaufort Force 6 may be expected. Within the circular cloud mass winds of F8 increase inwards, until normally at about 75 miles from the storm centre the wind speeds will have risen to F12 or more (at least 64 knots). These are 'steady' wind speeds or 'maximum sustained winds'; gusts will far exceed these limits. The strongest winds are found under the annular region of thick cloud and torrential rain. Here, maximum sustained winds not uncommonly exceed 100 knots in a mature hurricane. In some, maximum sustained speeds in excess of 130 knots have been estimated or recorded by aircraft reconnaissance: these, at least in the western North Pacific Ocean, have been designated Super-Typhoons; similar 'super-storms' probably occur in other regions. Within the eye, winds are light or calm though the sea state is chaotic due to the huge waves entering the eye from practically every direction.

Wind directions are anticlockwise in the northern hemisphere and clockwise in the southern. The air drawn into a hurricane spirals inwards in a near-horizontal fashion at first but with an increasing vertical component which reaches a maximum in the annular cloud mass towards the storm centre. Within the eye the air is in fact slowly descending, hence its almost cloud-free nature. The visibility within the inner regions of a hurricane is almost nil due to the constant spray and torrential rain.

As long as a hurricane remains over a warm sea it normally retains its violence. Some do not recurve on reaching the western sides of the oceans but instead move over the continental land masses where they will usually die out in a day or two as their supply of moisture (latent heat) is cut off. As these storms approach the coast, sea levels rise much beyond normal and considerable flood damage occurs. The currents produced by these boosted water levels as they build and recede are also phenomenal.

After recurvature, those hurricanes which survive into the temperate latitude belt are known as Extra-tropical Storms. Here their diameter has usualy increased and their winds have decreased though maximum sustained winds often still exceed Force

12. They often retain some semblance of an eye at this stage.

It is not unknown for such storms, having raged over the Caribbean and Gulf of Mexico, to have become extra-tropical and lashed the coasts of north-west Europe with hurricane force winds a week or so later. Similarly, typhoons have been known to survive a crossing of the whole North Pacific and to have caused considerable damage, as extra-tropical storms on the coasts of British Columbia, while Coral Sea hurricanes have crossed the South Pacific to decay a week or so later in the vicinity of Cape Horn.

Global Distribution and Seasonal Variation

In general the more mature tropical revolving storms occur over the central and western parts of the tropical oceans. This is largely due to sea surface temperatures on the eastern sides of the oceans being generally lower than the 27°C/80°F threshold necessary for storm formation. Ocean current circulations are such that cool water is brought equatorward on the eastern sides of the oceans (Diagram 75). There are, however, two areas of exception; one of the eastern side of the North Pacific off the west coast of Central America and the other off the west coast of Australia. In both areas the shape of the coastline and the pattern of the ocean currents combine to prevent cool water entering these regions and so sea surface temperatures in their respective summer seasons become high enough to generate tropical revolving storms.

Another important exception, now on the western side of the ocean, is the absence of storms in the South Atlantic. Here the warm west-going South Equatorial Current (Diagram 75) is diverted by the 'bulge' of Brazil to run northwest into the Caribbean so that sea surface temperatures in the tropical South Atlantic are barely high enough to generate tropical revolving storms, at least in the area beyond latitude 5°S.

The main areas in the northern hemisphere affected by tropical revolving storms are:
—the western North Atlantic, including the Caribbean and Gulf of Mexico
—the eastern North Pacific
—the western North Pacific, including the South China Sea
—the Arabian Sea and Bay of Bengal

and in the southern hemisphere,
—the South Indian Ocean
—the western South Pacific Ocean.

We shall now consider the behaviour and seasonal variation of storms within each area.

The Western North Atlantic (Hurricanes)

Diagram 60 shows several storm tracks in this area. It can be seen that most are formed within the latitude belt 10° to 20°N, and east of longitude 50°W, though some originate within the Caribbean. Again, most recurve in about longitude 80°W, though there is considerable variability in the longitude of recurvature, while a few cross the mainland coast and decay overland. All coasts within this area, including eastern states of the USA, are liable to be affected by hurricanes, but those of South and Central American northward to Costa Rica are only indirectly affected by heavy swell and strong winds around the periphery of hurricanes raging further north in the Caribbean.

The 'hurricane' season, in this area, begins late May and normally ends by early December. The period from January to mid-May can be reckoned to be storm-free. Yachts on passage westwards to the West Indies normally delay departure from the Canaries, Madeira or the Azores until early

60. Typical hurricane tracks – Western North Atlantic

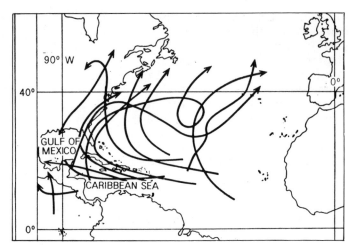

61. Typical hurricane tracks – Eastern North Pacific

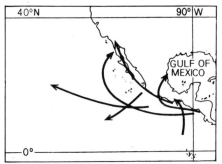

time, any part of the coast between Costa Rica and the Gulf of California, while others continue to run west-northwest to decay east of the Hawaiian Islands.

Vessels hauling north or northwest from Panama may be affected in some months, but those proceeding towards the South Pacific are unlikely to meet a tropical revolving storm in any month.

The tropical storm season on the western side of Central America is a little shorter than it is in the east. It begins in June and can be considered over by late November, during which time, on average, 14 tropical storms will have occurred (with a peak frequency of four in both August and September), several of which will have reached full hurricane maturity. On rare occasions a Caribbean hurricane crosses central America to rage over the eastern North Pacific.

Western North Pacific (Typhoons)
Most tropical revolving storms in this region originate near the Carolines in about 10° to 15°N, though some may form in about 5°N, and move west and later northwest before recurving north-eastwards to become extra-tropical depressions. Early in the season most recurve before the longitude of the Philippines, but as the season goes on

December so as to arrive at their destinations after the hurricane season is over, and usually leave before late May. Statistics based on the frequency of moderate tropical storms (Force 8 or more), any one of which may have reached the hurricane stage, show that they increase in frequency from June to reach a maximum of three or four in September in the average year. In the whole season, about ten tropical storms occur on average each year, six of which reach hurricane intensity.

The Eastern North Pacific (Hurricanes)
Due to the variability in the direction of the upper winds, which 'steer' tropical revolving storms, storm tracks are extremely variable, as shown in Diagram 61. Many follow a west-northwesterly track before swinging northwards to affect, from time to

recurvature takes place farther and farther west and so first these islands and the mainland coast of Asia and later the Japanese islands are increasingly liable to be affected. Some form in about 160° to 180°E and generally move north-northwest, though these rarely develop to full typhoon intensity. Others form in the South China Sea late in the season and again rarely reach full maturity. Some typhoons do not recurve but cross the mainland coast to decay in a day or two; of these, a few cross the Kra Isthmus and become reinvigorated, and now called Cyclones, in the Bay of Bengal.

Because sea temperatures in the western North Pacific, southward of about 15°N, generally remain above 27°C/80°F throughout the year, the

tropical storm season here, where 30 occur in a normal year, is much longer than elsewhere. The frequency is greatest from July to November with a peak in August when seven storms may develop in the average year, but they may also occur in the 'quiet' months from January to March when one storm may occur in each month every two years.

Any individual typhoon, but more especially those occurring from July to October, may reach super-typhoon intensity (maximum sustained winds over 130 knots). The circular cloud mass of such storms may well exceed 600 miles in diameter and the eye may be more than 30 miles wide. Similar 'super storms' probably occur elsewhere.

The Arabian Sea and Bay of Bengal (Cyclones)

Storm tracks in these areas show a very high degree of variability, though most tend to move in a north or north-westerly direction having formed in about 10° to 15°N in the central or eastern parts of these sea areas. However, an occasional Bay of Bengal storm has moved westwards to affect Ceylon while a few tropical storms in

63. Typical cyclone tracks – Arabian Sea and Bay of Bengal

62. Typical typhoon tracks – Western North Pacific

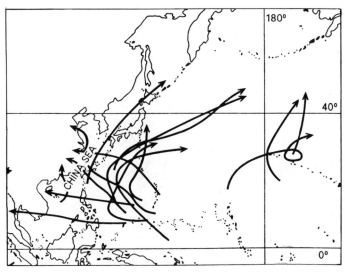

the Arabian Sea have moved westwards into the Gulf of Aden, later decaying over the African mainland. The storms of these regions, particularly the Bay of Bengal, can be as devastating as elsewhere; in November 1970 about 300,000 deaths were reported in Bangladesh when a severe cyclone crossed the coast at the head of the Bay of Bengal.

Cyclones are fairly infrequent in the Arabian Sea. In an average ten year period perhaps only 11 storms will occur and most of these will be in May, June, October and November. They are somewhat more frequent in the Bay of Bengal when 36 may be expected in an average ten-year period. Again there is a double 'peak', in May and October-November.

South Indian Ocean (Cyclones)

The tropical revolving storms of this region form between 5° and 15°S in a wide longitude belt stretching from 100°E (close to the Sunda Straits between Java and Sumatra) to about 45°E (near Madagascar). Those forming east of about 70°E normally travel southwest and later southeast before decaying southwards of about latitude 20°S while those forming further west tend to be more vigorous and longer-lasting. These latter cyclones sometimes affect the Mauritius, Reunion and Rodriguez island groups as well as Madagascar and the islands farther north, almost as far as the Seychelles.

The cyclone season in the South Indian Ocean lasts from December to April though some storms have occurred as early as October, with a peak spanning January and February, when two to three cyclones will occur in each of these months in the average year.

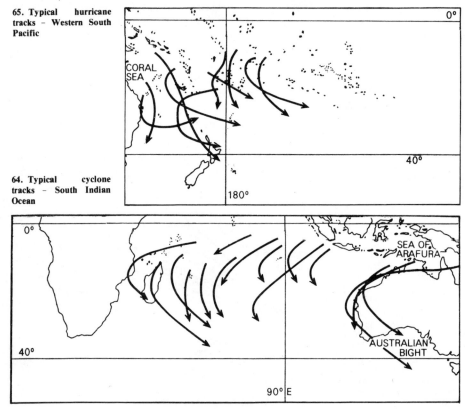

65. Typical hurricane tracks – Western South Pacific

64. Typical cyclone tracks – South Indian Ocean

The northern and western coasts of Australia are also affected by cyclonic storms. They normally develop off northern Australia and move west and later south around the Western Australian coast before most decay southward of the Great Australian Bight. On Christmas Eve 1974 a cyclone devastated the town of Darwin in Northern Australia.

The Western South Pacific Ocean (Hurricanes)

The storms of this region form over a wide longitude belt from the Coral Sea to about 160°W. Most develop in about 10° to 15°S and move in a southerly direction before turning southeastwards in about 20°S and later becoming extra-tropical depressions. The Queensland coast of Australia and New Zealand's North Island lie in the hurricane belt as do the islands of the Coral Sea and the island chains eastward to about 150°W.

In this southern hemisphere location also, the season lasts from December (sometimes November) to April, with a peak frequency in January when two cyclones will occur in an average year.

Surveillance and Warnings

The stations transmitting tropical revolving storm warnings are listed in the Admiralty List of Radio Signals Vols 3 and 3 (a) and similar publications from other maritime nations. Each area within the hurricane belt is covered by a responsible main meteorological office. These Storm Warning Centres are equipped with satellite picture receiving apparatus, both visual and infra-red, so that several pictures are received each day. In some areas such as the American sector pictures from geostationary satellites are received every 30 minutes (see Appendix 5).

It is fairly easy to detect tropical revolving storms from satellite pictures, as the area beyond the storm's feeder bands is normally an area of descending air and is therefore usually cloud-free (Plate 16): it is unlikely that any storms are going to evade detection by the intensive 'storm watch'. Indeed, the statistics of storm frequency in such areas as the Indian Ocean, where islands are generally few and far between and shipping has been none too frequent, and to a lesser extent in all other affected areas, will almost certainly be revised upwards when sufficient satellite data have been accumulated.

A study of satellite pictures over several years has revealed that a good estimate of the maximum sustained winds within a particular storm can be obtained from the diameter of its circular cloud mass: the larger this is, the stronger the winds.

In the more densely populated areas such as southeast Asia and the Caribbean, storm warning radars with a range of several hundred miles have been erected at several elevated sites so that each storm remains under constant surveillance. In these areas aircraft reconnaissance into each storm provides data on wind speed, cloud, temperature, diameter of eye, etc.

The past track and speed of a storm are, of course, easily obtained from the wealth of information, particularly from satellites, now available. Nevertheless each storm warning centre must also warn of its predicted path and speed over the next 24 hours or so. In formulating a storm warning, the past track will be considered, alongside other factors such as the upper air pattern over the general area, the present position of the storm and the time of year. The warning will contain details of the storm's position, its present state (i.e. moderate tropical storm, hurricane etc) amplified by present estimates of its maximum winds, and its predicted movement and development over the next 12 to 24 hours.

These warnings will be repeated every few hours.

Since tropical revolving storms often behave in a most erratic manner, more especially at recurvature than at other times, and since they may sometimes develop explosively from cloud cluster to hurricane within 24 hours, it is obviously in a vessel's best interests to constantly monitor the storm's movement for itself, as well as listening to all radar warnings.

Signs of Approach

Of course, small vessels such as yachts, fishing boats, tugs, etc should avoid the hurricane season if possible. In some areas, such as the western North Pacific, where storms may occur in any month, a listening watch should be kept for storm warnings and, concurrently, a weather eye kept lifted for these signs of the approach or formation of a tropical revolving storm.

1. A long swell (crest-to-crest period over 10 seconds) is usually the first indication of a storm in the general area. The swell will be running out from the approximate centre of the storm.

2. A fall in pressure of about 3 mb below the values at the same time over the past day or two is usually a good indicator of a tropical storm in the vicinity. It is necessary to compare pressures for the *same* time on consecutive days because, more especially within the tropics, there is a marked diurnal (twice daily) variation in pressure, like an atmospheric tide, which is characterized by peaks at about 1000 and 2200 and 'troughs' at about 0400 and 1700 local time, the diurnal range (rise and fall) being between 2 and 3 mb. (See Diagram 66.)

3. Steadily thickening cloud, at first in the form of cirrus but later becoming darker and more threatening, especially if this is first observed in the direction of approach of any long swell.

4. A freshening wind from a different direction from that which normally prevails over the area. (This is only appropriate in those areas where the winds are normally steady, such as the Trades areas or during the Southwest Monsoon of the South China Sea.)

These four factors, individually or better still collectively, warn of the existence of a storm in the area; spreading cloud and swell may give an indication of its bearing, otherwise there is no indication of its position and movement and these must be determined if the storm is to be avoided.

Avoidance of Tropical Revolving Storms

The best indicator of the position and movement of a storm (other than radio warnings) is the wind itself, especially if this is blowing from an unusual direction. Having suspected that a storm is in the vicinity, now apply Buys Ballots' Law (back to the

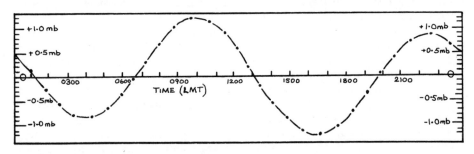

66. The diurnal variation of pressure within the tropics. The zero line is the average pressure for the locality, given in Sailing Directions etc. If this is unknown use the yacht's observed pressure at the same time yesterday to determine the real pressure change: if more than 3.0 mb then suspect that a tropical revolving storm is in the area.

wind: low pressure on left (right) in northern (southern) hemisphere). In fact while the wind remains no stronger than Force 6 the storm centre will lie about 70° on the left hand in the northern hemisphere, and to the right hand in the southern. When winds reach Force 8 or more, now within the umbrella of the circular cloud mass and with the barometer at least 5 mb below normal, the centre will lie directly on the left or right hand in the northern and southern hemispheres respectively.

Bearings, based on these assumptions, taken several hours apart, will give an indication of the storm's track (after allowance has been made for the vessel's movement).

Having established the storm's track, avoiding action may have to be taken. In this connection the terms Dangerous and Navigable Semicircles are often used; these apply to large vessels: the whole storm must be considered dangerous to yachts. These semicircles are separated by the storm's instantaneous track (Diagrams 67 and 68). To determine which semicircle the vessel is in (or about to enter), knowing the storm's track, if the winds are such as to blow the vessel into the projected

67. Rules for avoiding a hurricane – northern hemisphere	Dangerous Semicircle Yacht A Should bring wind on starboard quarter and progressively alter course to starboard as wind veers.	Ahead of storm Yacht B Should bring wind on starboard quarter and progressively alter to port as wind backs.	Navigable Semicircle Yacht C Should bring wind on starboard quarter and progressively alter to port as wind backs.

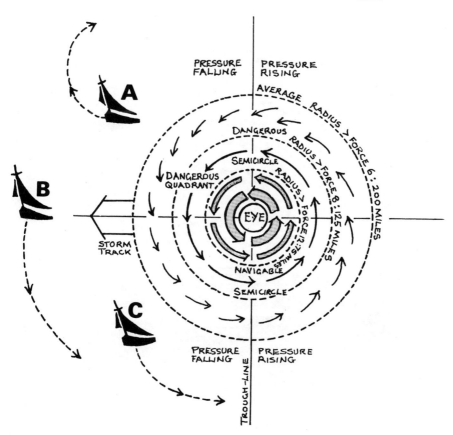

storms' track then the vessel is in the Dangerous Semicircle; if they will take the vessel *directly* into the area just ahead of the storm, then the vessel is in the Dangerous Quadrant. Conversely, if the winds are such as to blow the vessel towards the rear of the storm then the vessel lies in the Navigable Semicircle.

Or in another form:

—If, in the northern hemisphere, the wind is veering the vessel must be in the dangerous semicircle.

—If, in the southern hemisphere, the wind is backing the vessel must be in the dangerous semicircle.

—In both cases these rules always apply, irrespective of the storm's movement.

It is important to remember here that these storms sometimes backtrack, albeit infrequently, along or close to their original track, so that almost without warning the navigable semicircle becomes the dangerous one and vice versa.

Diagrams 67 and 68 give schematic representations of the wind field around a hurricane in the northern and southern hemispheres, respectively. They also show the dangerous semicircle and dangerous quadrant and the

| 68. Rules for avoiding a hurricane – southern hemisphere | Dangerous Semicircle: Yacht A bring wind on port quarter and progressively alter to port as wind backs. | Ahead of storm: Yacht B bring wind on port quarter and progressively alter to starboard as wind veers. | Navigable Semicircle: Yacht C bring wind on port quarter and progressively alter to starboard as wind veers. |

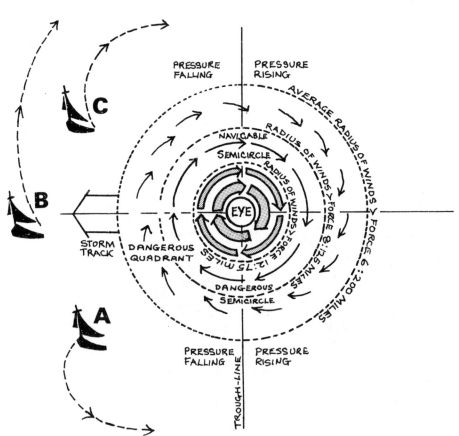

navigable semicircle relative to the storm track.

Vessels seriously threatened by a hurricane can be in any of three basic positions; either right ahead of the storm track or lying in either the dangerous or the navigable semicircle. Having determined which case applies the following avoiding action should be taken immediately.

Northern Hemisphere
A. A yacht in or about to enter the dangerous semicircle (wind veering) should broad-reach on starboard tack and maintain that point of sailing by progressively altering course to starboard as the wind veers until the pressure begins to rise when the storm will begin to move away (see 'trough line' on Diagram 67).
B. A yacht right ahead of the storm's track should maintain a broad reach on starboard tack by progressively altering to port as the wind backs, until the pressure begins to rise.
C. A yacht in or about to enter the navigable semicircle should also maintain a broad reach on starboard tack by progressively altering to port as the wind backs, until the pressure begins to rise.
Southern Hemisphere
A. A yacht in or about to enter the dangerous semicircle (wind backing) should broad-reach on port tack and maintain that point of sailing by altering to port as the wind backs.
B. A yacht right ahead of the storm should maintain a broad reach on port tack by altering to starboard as the wind veers.
C. A yacht in or about to enter the navigable semicircle should also maintain a broad reach on port tack by altering to starboard as the wind veers.

To summarize the 'rules' for detection and avoidance of tropical revolving storms:
Detection: from barometric pressure,
swell, unusual wind, high cloud increasing and persisting.
Position and movement: apply Buys Ballot's Law to determine bearing of storm. A second bearing should determine storm track.
Avoidance: Northern Hemisphere. Put wind on starboard quarter and keep it there by altering course to starboard if in dangerous semicircle or to port elsewhere.
Southern Hemisphere. Put wind on port quarter and keep it there by altering to port if in the dangerous semicircle or to starboard elsewhere.

But better still avoid the hurricane season, and if this is not possible then maintain a listening watch for tropical revolving storm warnings.

Tropical storms are given girls' names in many areas. In the western Pacific, the names of typhoons run alphabetically irrespective of calendar, while in the western Atlantic the names re-start with A at the beginning of each season. In this sector any hurricane which has caused considerable damage has its name removed from the list of names.

Waves

THOUGH waves are normally associated with the sea they form in the layer between any fluids of different density and also within the same fluid between layers of different density (this happens within the atmosphere and within the oceans). They are commonly initiated by frictional drag between one layer (fluid) and the other or by an obstacle in the flow, and they move in the direction of the stronger flow but at a fraction of its speed. The initial upward displacement in the wave, normally caused by drag, is compensated by a downward displacement due to gravity and this is repeated over and over again as long as the triggering action persists. They are called gravity waves.

To the scientist waves at sea are oscillations in the boundary layer between two fluids of differing density and speed (momentum). To the seaman they are at the least inconvenient, and at the most, terrifying.

It is common experience whether on lakeside or seaside or on the ocean, that there is a link between wind and waves: when the winds are strong the waves are big and when they are light or calm the sea or lake surface is generally smooth or almost so. As the wind increases slightly from calm, small ripples or catspaws appear on the

water surface which run quickly downwind, dying away or reforming as the wind comes and goes. With further freshening of the winds, the waves become larger and slow down. It might seem, at first, as though this would lead to a uniform pattern of waves arranged in rows with long crests and troughs at right angles to the wind. This might be the case if the wind increased everywhere over an area at exactly the same rate and from a constant direction. But this is not the nature of the wind; it is constantly varying in direction and speed so that its drag effect on the water surface is constantly varying and thus the surface becomes 'lumpy' with some waves higher than others and the crests often stretching for only a few metres or so across the wind. This is more or less the result of a wind of about 10 knots or so; at or below this speed the air flow (above and not at the boundary layer), though slightly variable, is laminar and reasonably smooth. At about 14 knots or so laminar flow within the air breaks down and at that speed and above eddies form within the airflow which strike down on the surface at irregular intervals to add to the lumpiness of the sea surface. As the wind increases further and the waves grow in size and move more quickly, a little more order seems to return and patterns of crests and troughs are more easily seen though continuous crests and troughs probably do not extend more than 100m or so across the wind. But here we must stop and consider the shape and the geometry of waves, and indicate some more precise relationships between winds and waves.

While the winds are fairly light, say below 12 knots or so, the waves take on the shape of a sine curve as shown in Diagram 69. The height of a wave (H) is the total distance from the top of the crest to the bottom of the trough. The wave length (L) is the distance between one crest and the next succeeding crest (it is *not* the length along the crest). The third parameter is the wave period (P) which is the time interval in seconds between one crest passing a fixed point and the next crest passing the same point; let C be the speed of a wave. The three parameters L, P and C can be mathematically linked through the following formulae to produce the relationships shown in Table 7.

$$C = 3.1 P \qquad L = 1.56P^2$$

Period (secs)	Length (metres)	Speed (knots)
2	6.2	6
5	39	15
10	156	31
15	351	47
20	624	62

Table 7. Values of wave length and speed corresponding to given wave periods.

There is no simple relationship between height (H) and the other three parameters.

As the wind increases above about 12 knots, the simple sine curve shape changes, as the upper parts of the wave become steeper, growing into the shape illustrated in Diagram 70, called a trochoidal wave form. According to theory the maximum 'stable' slope at the upper part of the wave is 1:2. At greater slopes the wave crest becomes unstable and breaks. To a very large extent this is borne out by experience.

To return to the general appearance of the sea surface with winds of say Force 5 (17 to 21 knots) which have produced a recognizable but far from continuous wave pattern, it is often the case that having reached a 'steady state' the wind direction changes substantially and soon generates another pattern of waves while the recent pattern is still running. These two patterns or wave trains may have almost identical height (in the simplest case) and speed etc, but their directions are appreciably different so that they either reinforce or cancel each other from time to time (Diagram 71). The resultant wave profile is much nearer to that which will be commonly experienced at sea. Further substantial wind shifts, perhaps with the passing of a depression, will produce further new wave trains to make the final pattern more complex, though in time the older wave trains will die away.

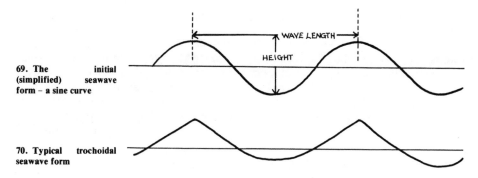

69. The initial (simplified) seawave form – a sine curve

70. Typical trochoidal seawave form

115

71. The resultant waves (lower diagram) from the addition of two wave trains of differing wavelength.

Initially the wave train heights are indentical (about 1.7 m). The resultant shows occasional waves of about 2 m (with some smaller ones) and maximum heights of about 3.5 m. The 2 m waves can be considered the 'significant waves'.

The diagram shows that two simple wave trains result in a much more complex pattern which more closely resembles that observed at sea (see Diagram 72). In many cases more than two wave trains are present.

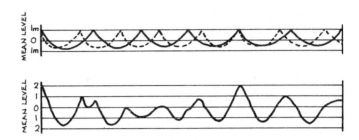

Waves associated with winds up to Force 6 or so (22 to 27 knots) normally die down within a few hours (say, 6 to 12) of the wind dying away. With persistent winds of Force 8 or more, normally generated by vigorous depressions, the large waves take much longer to die away - sometimes several days though during this time their height will have decreased from about 7 or 8m to about 1m and their wave length (and therefore speed - see Table 7) will have increased considerably. These waves are radiated from vigorous depressions into areas which may never become affected by the storm's strong winds.

Such wave-trains are called Swell and often indicate the presence of a storm, perhaps many hundreds of miles away. Swell is invariably present at sea and has to be added into the pattern produced by the wind to present an even more complicated picture. This is now more akin to the confused sea which sometimes sets vessels of all sizes pitching, yawing and rolling (and often bodily lifting or dropping them) seemingly all at once and, in extremis, evokes the question 'What am I doing here?' which most of us ask ourselves - time after time after time!

Despite the seemingly disorderly state of the sea surface there normally emerges a recognizable pattern of waves of almost similar height, length and period (though the wave crests and troughs themselves are normally shorter than 100m). These 'detectable' waves are called Significant Waves and are, by definition, the average of the highest one-third of the waves. They are in fact the waves which most observers would report as those being most representative of the state of sea being experienced.

Long wave-length swell moving rapidly from horizon to horizon may easily be detected from the bridge of a large vessel even where the wind-waves are well developed and the sea surface quite rough. In these conditions it is impossible to detect swell from the deck of a yacht, but when the wind-waves are small, say only 1m or so, swell may be felt from the yacht's motion rather than seen, especially the regular swell, with a period of between 10 and 15 seconds, radiating from a distant vigorous depression or tropical revolving storm.

The direction of this swell may be difficult to estimate from deck level but it should be determined if at all possible (perhaps by observation from the spreaders) for the generating storm will lie more or less in the direction from which the swell is coming. Since the storm may be located as far as a thousand miles away it may be that nothing more is experienced from it, but having noted the swell direction, both barometer and sky should be watched, even more closely, than before, for signs of the storm's

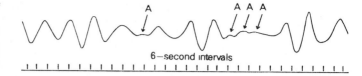

72. Typical ship-borne wave recorder trace

73. Relationship between wind speed, duration, fetch and the resultant wave height (after Dorrestein)

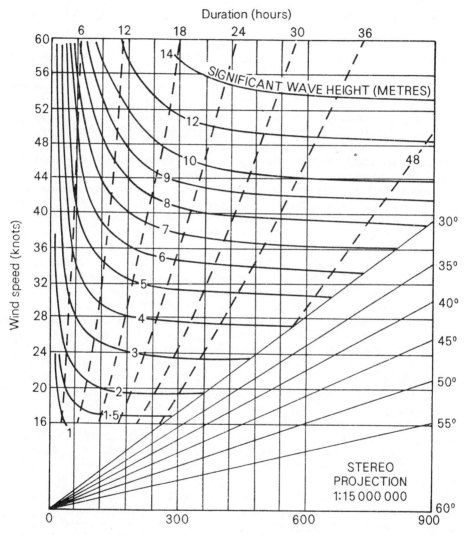

approach which will also be heralded by swell of greater height, even before the storm's wind-field affects the vicinity of the vessel. If an approaching storm (depression or hurricane) is suspected this is the time to consider avoidance action in accordance with Chapters 9 and 17.

Little has been said so far about wave height. In recent years wave recorders have been introduced on buoys and ships (Diagram 72). The relationships between their findings and the accompanying winds have led to some interesting results and show that wave heights are higher than those earlier estimated by eye for the same wind speed. This is hardly surprising due to the inherent difficulties in estimating wave height from a ship which itself is hardly ever straight and level.

It has long been known that wave height was dependent not only on the wind speed but also on its duration and fetch (the distance over open water up-wind over which the present wind has prevailed). The analysis of wind and wave recorder data over recent years has led to a relationship between the four variables (significant wave height, wind speed, duration and fetch) which is shown in Diagram 73. Knowing wind speed, duration and fetch, enter diagram with wind speed and then take the lower height of significant waves from fetch and duration. For example, wind speed 40 knots, duration 6 hours, fetch 300 miles gives results of 5m (6 hour duration)and 7.7m for 300 miles fetch; 5m would be chosen as the significant wave height. This table, based on wave recorder data, has been used to determine the new average wave heights given with the Beaufort Scale in Table 5.

There has been a considerable shift of opinion on maximum wave heights since the introduction of wave recorders, and also of offshore oil drilling rigs in areas of heavy weather such as the North Sea and off British Columbia. It is now generally accepted that waves of over 30 m (100 ft) can occur. These giant waves are most likely to occur on the equatorward side of a vigorous depression which is travelling rapidly (about 40 knots) from west to east. The larger waves in the westerly winds on the equatorward side of these depressions may well be travelling at depression speed, i.e. the gale or storm force westerly winds are continuously reinforcing the waves. Through the interaction of several

wave trains it is then possible for several crests from the various trains to come into phase and produce an isolated giant wave. These conditions are more likely to be met in high latitudes in winter and more especially on the eastern side of the northern hemisphere oceans, but almost anywhere in the higher latitudes beyond 40°S in the Southern Ocean.

Though waves may sometimes travel at considerable speed the water through which they propagate hardly moves at all, or at least there is little net displacement over the ground. This can easily be verified by observing the movements of an almost submerged floating object (little or no leeway) which will be seen to rise as a wave approaches, then to move *with* the wave at its crest, before descending in the rear of the crest and finally moving *against* the wave direction in the trough before repeating the process over and over again. Thus the float (representative of the surface water movements) describes circles in a vertical plane which is at right angles to the direction of the wave crests. These are not pure circles because the lateral (return) flow in the trough is a little less than that at the wave top so that surface water particles slowly move downwind (at a rate generally less than 1/10 knot).

Water particles beneath the surface describe similar near-circular paths but with decreasing diameter so that at a depth of half the wave length the movement is negligible. Huge waves generated by persistent gales such as in the Southern Ocean, which may be 10m or more high and whose length is about 200m, will not be felt at depths below 100m.

As waves reach shallow water (less than ½ wave length deep) the circular water particle movements become more and more elliptical and finally there is little or no vertical displacement. The movement is then towards the shore as the wave approaches and away as the wave passes by. This can be verified by observing the movements of seaweed in shallow water in fairly gentle conditions. The affect is also well known to swimmers who suffer sometimes strong 'under-tow' as a breaker passes by.

Due to the 'drag' from the bottom, waves decelerate as they approach the shore. The wave length is thus decreased but as their height is maintained they become steeper until the wave becomes unstable and breaks at its crest. In the same way the inshore ends of waves running obliquely towards the shore, or even perpendicular to it, are slowed by friction on the bottom and so the waves are bent (refracted) so that they finally run parallel or almost parallel to a straight gently shelving beach by the time they reach the breaker stage (Diagram 74).

In a similar way waves are refracted around a headland to run into a bay which otherwise offers a 'lee'. Again, waves may be refracted around islands to meet and produce a chaotic and potentially dangerous wave pattern in the lee of the island. A submerged bank (depth less than ½ wave length) also produces this effect and in rough conditions will produce dangerous seas in the lee of the bank (Diagram 74). The loss of a least one North Sea trawler has been attributed to this effect close to the Dogger Bank.

Waves breaking against a cliff, mole or similar steep shoreline rebound so that a 'reflected' wave may be seen to run seaward. This new wave-train interferes with the normal shoreward progression of waves, sometimes reinforcing it and at others reducing the wave height (in or out of phase) to make conditions more chaotic.

So far we have considered the propagation of waves running through otherwise still water. The situation becomes more complex when the water itself is moving over the ground under

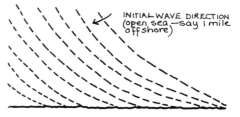

74a. Wave refraction on a straight beach

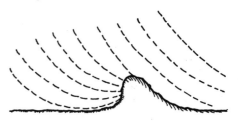

74b. Wave refraction around a headland

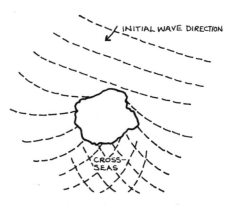

74c. Wave refraction around an island or shoal

tidal or current influences. This is the well-known 'wind against tide' effect. While the wind-waves are running with the tide sea conditions are compatible with the wind strength, but when the tide opposes the wind-waves the latter slow down and become steeper: it becomes more 'choppy'. The effect is so noticeable in some areas such as the Solent that the changed conditions indicate that the tide has turned. It also occurs in regions where the ocean currents are well-marked such as at the western sides of the oceans (see Chapter 19). Wind-waves generated by persistent strong winds of gale or near-gale force running against a strong current (say 3 knots or so) will decelerate below their still-water speed and so become steeper. It has been calculated that if the rate of the current or tide stream is ¼ of the still-water wave speed the waves will become unstable due to their increased steepness and the crests will break. A notorious area for this wind-against-current effect is off the east coast of South Africa, in the southern hemisphere winter, where several large ships have suffered serious damage and some have foundered. The Agulhas Current running southwest at up to 4 knots, under southwesterly gale conditions, opposes waves running northeast. Due to the heavy Southern Ocean swell being refracted into the South Indian Ocean the resultant waves in the region of the Cape Province and Natal coasts may attain heights of 20 to 30 m. The slopes of these waves, towards both the crests and the following troughs, will be unusually steep so that the crests break in a turmoil which would damage a large vessel and destroy a yacht. When making a southwest passage towards the Cape of Good Hope the best time to choose is the southern hemisphere summer when strong winds are much less frequent. The area should clearly be avoided in the southern winter when southwesterly gales occur; at this time a 'safer' passage may be made closer inshore, within the 200 m depth line where a weak counter-current runs northeastwards inshore of the Agulhas Current.

This dangerous wave effect may occur in any oceanic region where gale conditions oppose a strong current, and the Gulf Stream is a further example.

A last wave effect, which is commonly called a 'tidal wave' but has nothing to do with the 'wave' which, due to the gravitational attraction of the moon and the sun, circumnavigates the globe twice each day, is that due to coastal and, more especially, submarine earthquakes - a Tsunami. These waves, caused by vibrations in the earth's crust, have extremely long wave lengths (commonly several hundred kilometres) and travel at enormous speeds which are of the order of several hundred knots. Their height over the open ocean may be half a metre or less and they are quite undetectable here. However, on approaching a coast these waves decelerate and may sometimes cause the sea to rise 10 to 20m above its normal level with consequent loss of life and damage to property. Tsunamis (the name is Japanese) are more frequent in the Pacific Ocean than elsewhere due to its volcanic perimeter.

In estimating wind speed from the state of the sea surface, allowance must be made for the considerable sheltering effect from a nearby coast; this may extend for several miles offshore. The wind speed should have been steady for at least several hours so that the seas are fully developed. A recent change of wind speed, say within the past half-hour or so, will have had little effect on the sea state except in the circumstances, described above, where large waves may be travelling at depression speed and therefore arrive simultaneously with the stronger winds.

CHAPTER 19

Ocean Currents

BY ocean currents is meant the movement of water due to non-tidal influences. Near most coasts and within the 200m contour, the edge of the Continental Shelf, tidal streams tend to dominate the water movements. Further offshore the tidal effect is negligible, though in these oceanic regions the water movements are sometimes considerable. By convention, current directions or 'sets' are those *towards* which the current is moving, i.e. a westerly current or set is a west-going current; this is *opposite* to the convention for naming wind directions.

The ocean is a fluid which is never at rest. Within the ocean there is a whole spectrum of eddy sizes, varying from the giant oceanic circulation (Diagram 75) to small circulation of only a few miles diameter, or sometimes even much less. As within the atmosphere, water particles may move in three dimensions due to the influence of wind stress on the surface and to pressure differences on and beneath the surface which are produced by horizontal density differences as well as slopes in the sea surface over several hundred miles produced by varying wind stress.

The wind effect is said to produce directly a wind drift current (due to drag) which sets downwind but is off-set by 20° to 30° to the right in the northern hemisphere and to the left in the southern, at a rate which is proportional to the wind speed, duration and fetch. Just how this comes about, when most of the wind energy is transferred into producing and maintaining waves in the sea surface, is not easily understood. We have already seen (Chapter 18) that wave action produces a slight net transport of water downwind, albeit at a slow rate. Nevertheless with persistently strong winds as in a slow-moving depression or better still the Trades, vast quantities of surface water are transported in this fashion which would probably result in a sloping sea surface over the area affected by the persistent wind. This slope (upwards from the upwind end to the downwind end), as we shall later see, would not produce a current setting in a down-wind direction. It may well be that the wind effect in reality is a 'slope-producing' effect, piling-up the water before it.

For this reason and also due to density differences within the oceans, sloping surfaces of constant water pressure (compare upper air charts showing contours of constant atmospheric pressure in Apppendix 4) are not uncommon in the oceans. Sea water tends to move down the slope but is deflected by the earth's rotation towards the

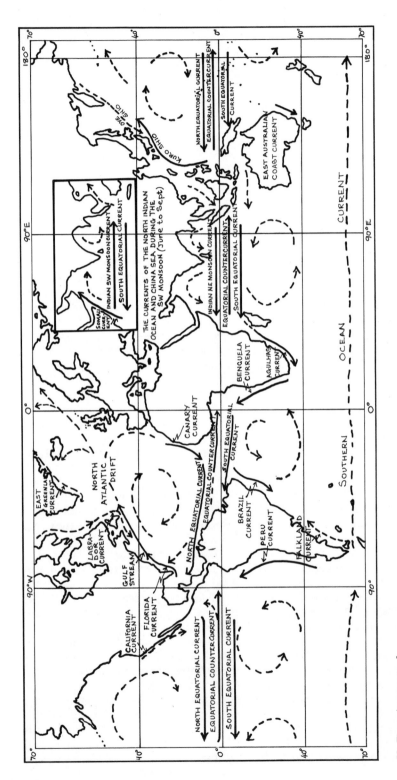

75. The main currents of
the oceans

right (left) in the northern (southern) hemisphere and finally runs in a balanced state 'across' the slope just like the wind. The resultant water movements are known as Gradient Currents.

Ship observations of currents, gathered over many, many years, clearly indicate that over the greater part of the oceans the currents are variable and generally slow, but they also indicate that in a relatively few areas the currents run at a considerable rate in a more or less constant direction throughout the year or, less commonly, throughout a season. The main regions so affected are the equatorial regions and the western and, to a lesser extent, the eastern boundary zones of the oceans.

It is conceivable that the obliquely converging trade winds produce a piling-up of water along the whole length of the equatorial region. Due to the negligible geostrophic effect in this near equatorial region (see Chapter 17) gradient currents cannot be produced, neither can the water run straight down the slope on either side as the wind (wave) effect is constantly bringing more water into this region. The only way water can escape is in a general westerly direction, i.e. downwind. Thus on both sides of the equator there are west-going currents. However these are separated by an east-going equatorial countercurrent. The reason for this easterly current is not clearly understood; it is probably a result of a complex three-dimensional ocean circulation in the equatorial region. Nevertheless, the fact that these three equatorial currents exist is beyond dispute.

Leaving aside the countercurrent meantime, the west-going equatorial currents are deflected by the continental shelves on the western side of the oceans to run, often strongly, polewards. These currents are often referred to as the Western Boundary Currents. They then break away from the continental coasts and fan out in a general easterly direction, later gathering momentum as they return equatorwards on the eastern sides of the ocean before finally turning westwards to complete large oceanic circulations which are clockwise in the northern and anticlockwise in the southern hemisphere. The main currents of these circulations are named in Diagram 75. Ships and yachts on ocean passages, of course, plan to use or avoid these currents. Though even those which are described as persistent show some variability in direction, sometimes temporarily opposing the mean flow. On a long passage there will be a considerable benefit from using them or by staying clear, as appropriate.

The North and South Equatorial Currents in each ocean run strongly westwards at mean rates of about 1 to 2 knots and sometimes 3 to 4 knots. These currents are separated by an east-setting Equatorial Countercurrent with similar rates. In general, the countercurrent lies just north of the equator in the Pacific and Atlantic Oceans; the North Equatorial Current lies between about 5° and 10°N and South Equatorial Current just south of the equator. There is a small seasonal shift in these general boundaries of about 1° in latitude north and south following the sun.

However, there are several exceptions to this general description. Though it applies well enough in the Pacific Ocean, the picture is modified in the Atlantic. There the South Equatorial Current is in part deflected across the equator by the coast of Brazil. Also, the countercurrent does not prevail in all longitudes; in fact it varies seasonally from a maximum development in autumn when its source can be traced in about 50° W, to a minimum in spring when the source is in about 20° W.

A much greater exception occurs in

the Arabian Sea in the Bay of Bengal and in the South China Sea, where the current pattern is completely reversed by the changing monsoons. During the Northeast Monsoon the currents closely resemble those of the other oceans except that the countercurrent lies just south of the equator. During the Southwest Monsoon the northern boundary of the countercurrent merges with a general 'easterly' set in the Arabian Sea and Bay of Bengal (see Diagram 75).

The other main ocean currents are those on the western and, generally to a lesser extent, on the eastern sides of the oceans. The main western boundary currents are the Gulf Stream in the North Atlantic, the Kuro Shiwo in the North Pacific, the Agulhas in the South Indian Ocean and, in the Southwest Monsoon only, the Somali Current in the Arabian Sea. All of these currents set polewards off the edge of the continental shelf (200m contour). The axis of strongest flow at the surface lies within a few miles of the 200m depth contour and the current gradually weakens further offshore. They run at a considerable rate which often maintains 2 to 3 knots on average and sometimes more, and the maximum may reach 7 knots. (Yachts using these strong currents, especially in winter, should be aware of the 'wave against current' effect referred to in Chapter 18. To avoid these regions when strong opposing winds are expected, a detour inside the 200m contour will take the vessel away from the dangerous seas.) The Brazil Current and the East Australian Coast Current run at rates much weaker than the other western boundary currents.

All of these western boundary currents turn towards the east in about 30° to 40° latitude, where they begin to lose momentum as they run across the oceans. In the northern hemisphere a part of these set more northeasterly into high latitudes while the remainder turn southwards towards the equator on the eastern sides of the oceans. In the southern hemisphere a branch joins the West Wind Drift which encircles the globe while the remainder turns north towards the equator. In general these Eastern Boundary Currents are weaker than their western counterparts. For example the Californian and Canaries Currents run at a rate of about 1 knot, sometimes 2 to 3 knots; however, the Benguela and Peru Currents, off the west coasts of South Africa and South America respectively, are stronger and run at rates of 1 to 2 (sometimes 3 to 4) knots.

All of these currents turn westwards as they approach the equator and thus complete the oceanic circulations.

One western boundary current worthy of a closer inspection is the Gulf Stream, which has a considerable influence on the climate of northwest Europe even well north of the Arctic Circle (see Chapter 20), warming it in winter and cooling it in summer. The source of the Gulf Stream is the Gulf of Mexico where a head of water results from the Equatorial Currents flooding westwards through the Caribbean. Its only outlet is the Florida Strait through which it flows at rates of 4 to 5 (sometimes 6 to 7) knots, as it rounds the Florida Peninsula; here it is known as the Florida Current. North of about 28°N the main stream is joined by a minor current which has set northwest outside of the West Indies, the Antilles Current; the combined flow, now known as the Gulf Stream, follows the edge of the continental shelf to Cape Hatteras and then proceeds more northeastwards towards the tail of the Grand Banks off Newfoundland.

As the current swings away from the continental shelf near Cape Hatteras, it begins to oscillate gently in a horizontal plane. The oscillations often develop into meanders and some of these finally become complete cut-off

1 Cirrus

2 Cirrostratus (with altocumulus patches at lower left)

3 Altostratus

4 Nimbostratus

Plates 1 to 4 represent the typical cloud sequence as a depression (especially its warm front) approaches – see Chapter 9. Cirrus clouds spread from some westerly point and thicken to cirrostratus, through which a halo may be observed around the sun or moon. After a few hours the cirrostratus thickens to altostratus (which almost obscures the sun) and later to nimbostratus (which completely obscures the sun) from which continuous and often heavy rain falls. Low stratus normally forms in the rain at only a few hundred metres above the earth's surface. On some occasions these cloud developments may be hidden behind a complete cover of lower cloud.

5 Small cumulus (*cumulus humulis*)

6 Medium-sized cumulus (*cumulus mediocris*)

7 Large cumulus (*cumulus congestus*)

8 Cumulonimbus

Plates 5 to 8 show the complete range of
convective cloud development from the
initial small cumulus stage to that of
cumulonimbus (Diagram 29 and Chapter
12). The complete process may sometimes
occur in less than 30 minutes. Plate 8
shows a typical mature cumulonimbus
cloud photographed from about 15 miles
(cloud top about 10 km). The lower part
of such a cloud is often obscured by a
'ring' of cumulus. The upper part
displays here the characteristic 'anvil'
which forms when convective 'bubbles'
reach the Tropopause and spread out.
The 'anvil' is composed of ice crystals:
note the 'cirrus' appearance of its edges.

9 Stratus

10 Stratocumulus

11 Altocumulus

12 Cirrocumulus

13 Geostationary satellite picture of the 'Old World' hemisphere taken on 22 April 1978 at 1225 GMT by Meteosat over the equator on the Greenwich Meridian (see Appendix 5). (Crown photo)

Plates 13 and 14 were taken on dates fairly close to the equinoxes at local noon over the picture centre (the satellite sub-point) and therefore the 'full' hemisphere is illuminated. The distortion due to the curvature of the earth's surface becomes increasingly apparent at about 50° latitude or longitude from the satellite sub-point.

The main points of interest are:

1. The mainly cloudy 'disturbed westerlies' belt beyond about 45° lat.

2. The mainly clear sub-tropical areas.

3. The circulation around depressions in both hemispheres (anticlockwise in N, clockwise in S).

14 Geostationary satellite picture of the 'New World' hemisphere taken on 3 September 1978 at 1701Z over the equator at 75°W (parallels and meridians shown at 10° intervals). (Photo courtesy NOAA)

4. Two hurricanes on Plate 14, 'Ella' off the east coast of the USA and 'Norman' off the west coast of Mexico.

5. The cloud cluster centred in the Atlantic at about 11°N, 38°W has every appearance of developing into another hurricane.

6. The 'band' of cloud at about 10°N in the Pacific and closer to the equator in the Atlantic is called the Inter-tropical Convergence Zone, the boundary region between the converging NE and SE Trades.

15 Satellite picture of hurricane 'Ella'. This is an enlargement from Plate 14. The mainly circular cloud mass, representing Force 8 or stronger winds, is about 300 miles in diameter; the area swept by the spiral bands (Force 6 or more winds) is about 500 to 600 miles in diameter; while the eye of the hurricane is about 15 to 20 miles across. Note the mainly clear area surrounding the hurricane. (Photo courtesy NOAA)

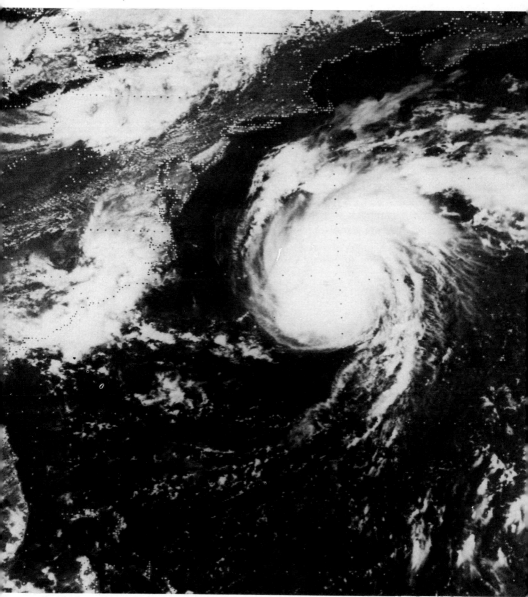

16 Satellite picture of a wave depression in the eastern North Pacific; North America is outlined top right. (Photo courtesy NOAA)

The picture is taken from a geostationary satellite centred over the equator at long. 135°W (the satellite sub-point) at 1845 GMT on 4 Feb 1979. This is 0945 local time at the satellite sub-point and so the northwestern point of the picture is still in darkness; the earth's terminator can be seen to be lying from about the Gulf of Alaska towards Hawaii.

A mature depression is centred in the Gulf of Alaska; its warm front reaches southeast along the west coast of USA, while its cold front runs southwestwards from the centre of the low. A developing wave depression can be seen on this cold front in about 30°N, 160°W (see Chapter 3). The 'solid' mass of cloud within the warm sector of the main depression is probably sea fog or low stratus, while the cloud in the lower part of the photograph is associated with the Inter-tropical Convergence Zone.

17 Satellite picture of meanders in the
Gulf Stream. (Photo courtesy NOAA.)
This infra-red photo (dark shades are
warm surfaces, lighter shades cool ones)
shows a marked boundary between the
warm (black) Gulf Stream at left centre
and the cooler inshore water; meanders
can be seen at this boundary. Earlier
meanders have become cut off on the
north (upper) side of the boundary and
can be seen as warm clockwise eddies
within the cooler water. Though the main
temperature contrast occurs at the edge of
the Gulf Stream, a second discontinuity is
seen between the cool waters on the
continental slope (slope water) and the
cold water over the continental shelf
(shelf water). The coastline is at upper
left, and the Great Lakes are in the
corner; white patches are cloud. There is
considerable distortion beyond the thin
vertical lines.

18 Satellite picture of the 'Fastnet
depression'. The photo was taken from
the polar-orbiting satellite TIROS N at
1535 GMT on Monday 13 Aug 1979. The
position of the low at this time ($50\frac{1}{2}°$N,
16°W) and its fronts have been
superimposed to aid interpretation. The
'clear' area immediately to the west and
southwest of the storm's centre was to
remain a feature of this storm and
probably accounts for several reports of a
clear 'eye' during the height of the storm.
The cloud structure ahead of the low is
confused by that of an earlier frontal
system which was then still affecting the
Southwest Approaches. A detailed
analysis of this storm is in Chapter 21.
(Crown photo)

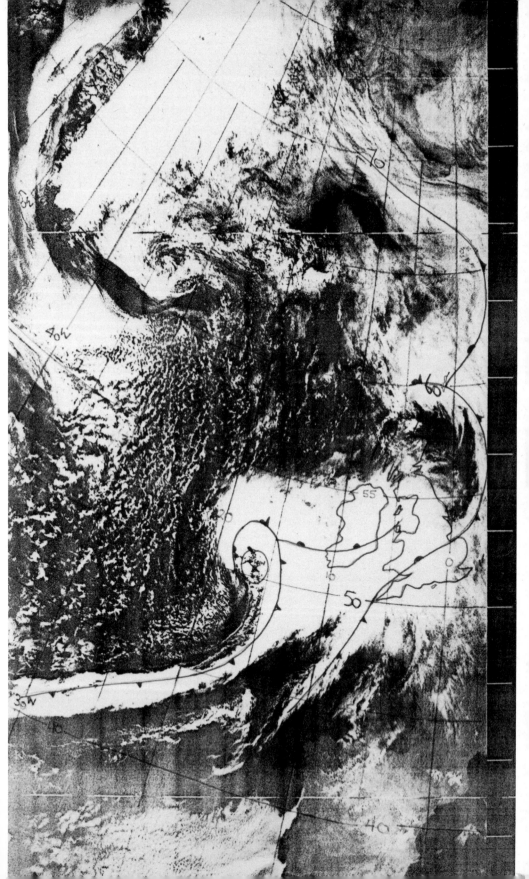

19 Aerial view of a mature waterspout.
The formation process is described in
Chapter 7. This well-developed
waterspout is estimated to be about 50
metres wide at its base. The disturbed sea
surface immediately below the 'spout is
clearly seen. Smoke floats in the
foreground and perhaps a few hundred
metres from the 'spout indicate the
'prevailing' surface wind direction beyond
the waterspout's narrow influence. (Photo
courtesy J. H. Golden)

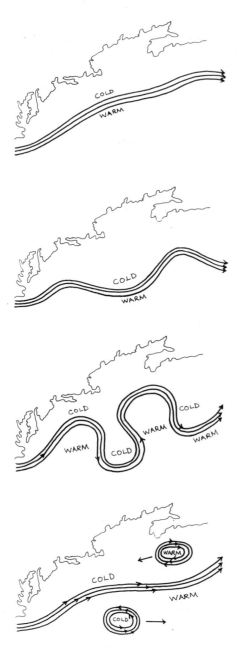

76. The development of meanders and eddies in the Gulf Stream

river. The resultant eddies, warm on the northwestern side and cold on the southeastern side of the Gulf Stream, set southwest and northeast respectively at a slow rate before decaying in a week or two.

Inshore of the Gulf Stream, in the region between the Grand Banks and Cape Hatteras, the southwest setting Labrador Current of Arctic origin brings cold water alongside the warm Gulf Stream. The two water masses do not mix and an oceanic front results which is often referred to as the North Wall (of the Gulf Stream). This front is so well marked at times that a temperature difference of 12°C/22°F has been measured in winter between bow and stern by a US Coast Guard cutter lying stopped across it.

Eastwards of the Grand Banks the main Gulf Stream current begins to fan out and loses most of its momentum. In the region between Newfoundland and the Norwegian Sea it is known as the North Atlantic Drift. A part of this current sets north towards Iceland while a major part continues northeast into the Barents Sea where it normally contains the advance of the ice edge in about 75°N throughout the winter (see Chapter 20). Another main branch breaks away from the southern flank of the main stream to set southeast, then south and finally southwest, now as the Canary Current, off the west coast of Africa, before it joins the North Equatorial Current thus completing the main current circulation within the North Atlantic Ocean.

Within this circulation, as in those of the other oceans, the currents are generally light and variable; however, they may well attain a considerable rate in the channels within island groups. This is particularly so where island groups lie within a strong current, such as the Maldives within the equatorial currents of the Indian Ocean. A yacht approaching an island group, even in a region of otherwise weak currents,

circulations, perhaps 100 miles across (see Diagram 76 and the Plates) somewhat in the same way as ox-bow lakes are formed from meanders in a

should do so with great caution as extremely strong currents may run in the narrow channels between the islands.

Sea Surface Temperature

We have already seen the importance of sea surface temperature as a factor in the formation of both tropical revolving storms and sea fog. We shall now look at the global patterns of sea surface temperature in February and August as being representative of the winter and summer conditions respectively in the northern and southern hemisphere.

Not surprisingly, there is a decrease of temperature with increasing latitude but the pattern is not quite that simple. Within about 40° of the equator there are higher temperatures on the western sides of the oceans than on the eastern sides due to the current circulations

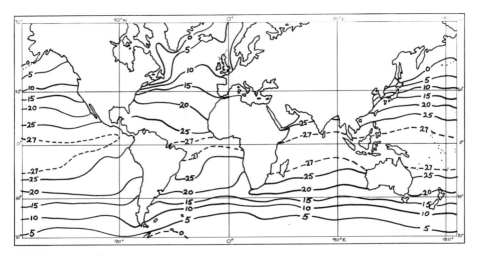

77. Mean sea surface temperatures (°C) – February (winter)

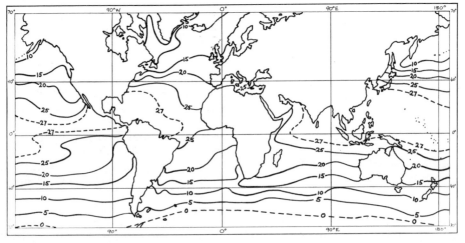

78. Mean sea surface temperatures (°C) – August (summer)

126

(see Diagram 75). Beyond 40° latitude, in the northern hemisphere, there is warmer water on the eastern sides of the oceans than in the west, again due to the currents. This effect is more marked in winter when colder water is brought equatorwards from the polar regions by the Labrador Current in the North Atlantic and, to a lesser extent, by the Falkland Current in the South Atlantic and again the Oyo Shio in the North Pacific Ocean.

To indicate the relevance of a critical temperature in the formation of tropical revolving storms, the 27°C isotherm has been included in Diagrams 77 and 78. When allowance has been made for the 'latitudinal' effect within 5° latitude of the equator in preventing any powerful air circulations, it can be seen from a comparison of Diagrams 77 and 78 with Diagrams 60 to 65 that there is a clear relationship between sea surface temperatures of 27°C or more and the breeding area of hurricanes, typhoons etc.

Where sea surface temperatures are unusually low for a particular latitude, sea fog is likely. There is no better example of this than the Newfoundland Banks region where the prevailing moist southwesterly winds flow over the cold water of the Labrador Current. The Californian coast is another area similarly affected.

Yet another important effect is that of upwelling, which produces sea fog off the coasts of Somali and Saudi Arabia during the Southwest Monsoon and also off the coast of southwest Africa and in some other areas. This arises from the fact that the prevailing winds in these regions have an offshore component which slowly moves the surface water offshore to be replaced inshore by sub-surface water. Since the temperature within the oceans decreases with depth (it is only a few degrees above freezing at the bottom of the oceans) this results in unusually cold water just offshore. It follows that a moist airstream flowing over this colder water may well produce sea fog.

The idealized sea surface temperature pattern (warm near the equator, steadily lowering polewards) is considerably modified by the ocean currents. We shall see in the next chapter that, for the same reason, currents also have a considerable influence on the distribution of sea ice in the polar regions.

CHAPTER 20

Ice

ICE at sea is not completely confined to the polar regions; in fact, in winter it is found surprisingly far from the polar regions in some longitudes. For example, in winter, sea ice normally covers the Gulf of St Lawrence and the Canadian coast north from the Belle Isle Strait, and in the Pacific it extends southwards to the north coasts of Hokkaido, the northernmost island of Japan. Sea ice also affects the northern portions of the Black Sea. In the southern hemisphere, in winter, the Antarctic sea ice normally extends to between latitudes 65° and 55°S, thus preventing great circle sailing between the Capes to anyone bold enough to attempt it in a small vessel. Even in summer, sailing in high latitudes in the southern ocean southward of about 55°S is hazardous due to large numbers of icebergs. On the other hand several coasts within the polar regions become accessible in summer.

There are two main types of ice to be found at sea. The first and overwhelmingly the more dominant is sea ice which results from the freezing of the sea surface. The second consists of icebergs, which are large pieces of land ice broken away from glaciers and ice-sheets, etc.

Salt water of average salinity (35 parts per thousand) freezes when the temperature, throughout its depth, falls to around minus 2°C/28°F. In the cold polar regions sea ice may attain a thickness of about 2 m during one winter season. In the depth of winter, well within the Arctic, where air temperatures may be as low as minus 30° to minus 40°C, an area of open water, formed when ice floes part or crack due to internal pressures, will normally freeze over to a depth of about 10 cm in 24 hours and 18 cm in 48 hours; these represent the maximum rates of ice formation.

As sea ice thickens its rate of growth decreases because the ice itself acts as an insulator between the very cold air above and the near-freezing water below. For this reason the maximum thickness of level sea ice, even that which has survived through numerous summers (and considerable melting occurs even within the Arctic each summer), is about 3.5 m, of which 3 m will be submerged. But the sea ice is not everywhere level; it is often heavily deformed into hummocks and ridges whose weight is supported by the buoyancy of vast quantities of ice under water. Maximum thickness of these deformed ice areas are probably about 55 m, of which about 10 m will be above sea level.

Sea ice is divided by its mobility into two classes. The one is pack ice consisting of pieces called ice floes, of varying size, thickness and age, sometimes a few miles across, reasonably free to move under the influence of wind and current. The other is fast ice which is immobile. Fast ice is found on the coasts outward to about the 50 m depth contour though where there are offlying islands the fast ice may extend much further offshore. In any case it is found only in coastal locations. Pack ice is, therefore, the dominant type found at sea.

Pack ice, where it prevails, does not everywhere cover the whole sea surface. That fraction covered by pack ice is called the 'concentration'; it is normally expressed in tenths. In this connection several terms will have to be defined:

—Ice Free means no sea ice present.
—Open Water means less than 1/10 concentration.
—Very Open Pack Ice means 1/10 to 3/10 concentration.
—Open Pack Ice means 4/10 to 6/10 concentration.
—Close Pack Ice means 7/10 to 8/10 concentration.
—Very Close Pack Ice means 9/10 to just less than 10/10 concentration.
—Consolidated Pack Ice means 10/10 concentration, i.e. no breaks. (This normally occurs when pack ice becomes jammed in channels between islands, such as in the Canadian Arctic, and the spaces between the floes become completely frozen over.)

The concentration of the pack ice normally decreases in the last few miles, or even less, towards the outer (oceanic) edge of the ice, but this decrease depends on the lately prevailing winds over a few days. Should these be from open sea onto the ice, then the ice edge will be well defined, probably as close or very close concentration (i.e. an immediate change from open sea to 7/10 to 9/10). If the winds have been 'off ice' then there will be a gradual reduction of concentration so that the ice edge is ill-defined. A hundred miles or more may then separate high concentrations of ice (say more than 6/10) from the outer edge which will be very open pack ice.

The ice regimes of the Arctic and Antarctic are greatly different. The Arctic is an ocean about 3000 m deep which is covered by a thin shell of sea ice, on average about 3.5 m thick, whereas Antarctica, of similar area, is a continent covered by an ice cap (compressed snow) which is up to 3000 m thick. The annual mean temperature at the South Pole is minus 49°C (the lowest temperature yet recorded, in winter, is minus 88.3°C), whereas at the North Pole the annual mean temperature is about minus 20°C (the lowest temperature yet recorded is only a little below minus 50°C).

The ice cap covering the Antarctic continent accounts for more than 90 per cent of the world's permanent ice; practically all of the remainder lies in the Greenland ice cap. (It has been calculated that the melting of these ice caps would raise the sea level globally by about 60m.) The ice constituting these ice caps is constantly moving outwards towards the coasts where many thousands of icebergs are calved each year from the glaciers and ice shelves which reach out over the sea. As a consequence of the inequality of size of the respective ice caps, large numbers of icebergs are to be found in a wide belt around the whole Antarctic continent, whereas the Arctic icebergs are to be found along the east and west coast of Greenland and, since they are taken there by the currents, along the eastern seaboard of Canada. Most of the Arctic Basin remains covered with pack ice throughout the year whereas the greater part of the Antarctic sea ice melts each summer.

Distribution and Seasonal Variation of Ice

Sea ice normally reaches its greatest extent in March and April in the northern hemisphere and least in August and September. In the southern hemisphere the greatest extent occurs in September and October and least in February and March: these times of greatest and least extent will be referred to as the winter and summer conditions respectively in each hemisphere. The advance of the ice edge from summer to winter takes place at a more or less even rate whereas the retreat from winter to summer begins slowly at first and reaches its greatest rate during the period from the solstice month to the end of summer when air temperatures fall dramatically in the polar regions.

Northern Hemisphere

Diagram 79 displays the mean position of the 4/10 ice edge in 'winter' and 'summer'. To show the variability from year to year the 1/10 maximum ice edge and the 7/10 minimum edge are also given as indicative of those areas which are sometimes affected by ice and those which are always affected by ice, respectively.

The effect of the ocean currents can clearly be seen on the shape of the ice edge, especially in winter, which lies more northeast to southwest than east to west (see Diagrams 79 and 80). In winter sea ice normally reaches as far south as 47°N (exceptionally 45°N) on the western side of the Atlantic under the influence of the cold Labrador Current, and to 44°N (exceptionally 37°N) on the western side of the Pacific Ocean. The effect of the North Atlantic Drift (originating as the Gulf Stream) and its branches is to maintain ice-free conditions to 76°N off southwest Spitzbergen where, despite extremely cold air temperatures, relatively warm sea water at the surface and at depth constantly flows into the

region. The greater part of the Norwegian and Barents Seas are also normally kept ice-free by this warm current, whereas 15° further south, in the inner Baltic, the Gulfs of Bothnia, Finland and Riga are normally ice covered. (Diagram 80 shows the boundary between the warm and cold currents: it can be seen that this lies very close to the 1/10 maximum limit shown in Diagram 79.)

As summer advances the ice edge retreats northwards at an increasing rate so that by August/September the original area of sea ice will have been reduced by about 25 per cent. Melting is achieved in two different ways. First, the ice is being constantly brought southwards by the cold currents into regions where the water is seasonally becoming warmer so that melting takes place farther and farther north. The other method, perhaps equally important, is by the melting of the snow cover which falls on the ice each winter. Though the snow is highly reflective, some melting will have occurred by June, due to constant daylight and often constant sunshine, resulting in puddles of water on the uneven snow/ice surface. These puddles absorb much more heat from the sun than the snow and ice do and so steadily melt the underlying snow/ice. This process of puddling achieves its maximum melting effect in late July when level sea ice, initially about 1.5 m in total thickness, may possibly melt completely and old, tough, level ice may be reduced in thickness from about 3.5m to 2.5m and will be considerably weakened.

As a result of summer melting, the sea ice in a normal summer eventually retreats within the Arctic Basin though a tongue of ice still stretches southwards along the east Greenland coast to about 70°N and many channels within the Canadian Arctic remain obstructed.

The minimum 7/10 limit would seem

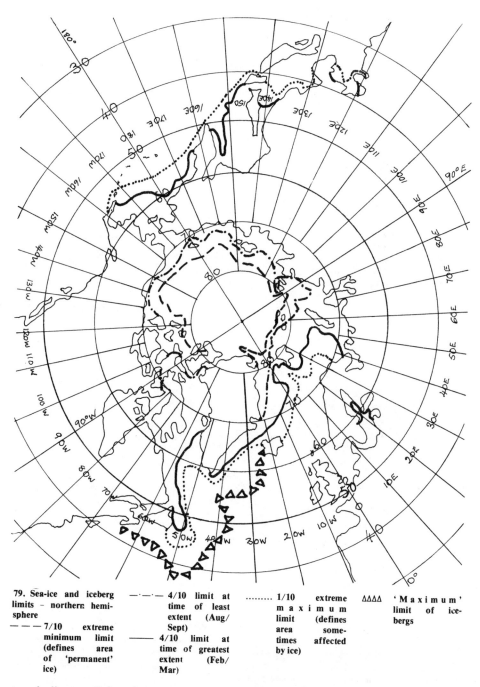

79. Sea-ice and iceberg limits - northern hemisphere

--- 7/10 extreme minimum limit (defines area of 'permanent' ice)

—·—·— 4/10 limit at time of least extent (Aug/Sept)

——— 4/10 limit at time of greatest extent (Feb/Mar)

········· 1/10 extreme maximum limit (defines area sometimes affected by ice)

ΔΔΔΔ 'Maximum' limit of icebergs

to indicate light ice or ice-free conditions within the Canadian Arctic Archipelago (the Northwest Passage) and along practically the whole northern coast of Russia (the Northern Sea Route) in exceptional summers. This is *not* the case. This 7/10 minimum limit is an envelope of the minimum conditions in each area in any year. It is a fact that very light ice conditions in

80. Surface currents in the Arctic Basin and North Atlantic. Broken line shows approximate boundary between 'warm' and 'cold' currents.

summer within one sector coincide with heavy conditions in another, sometimes adjacent, sector. For example, heavy ice conditions off the east coast of Canada normally occur simultaneously with light conditions over the Greenland and Barents Seas. This is because the winds, as well as the currents, have a considerable influence on the position of the ice edge. Warm southerly winds (over a week, or better still several weeks and exceptionally a season) drive the ice northwards and also melt it, even in winter, whereas cold northerly winds produce the opposite effect. It is clear that persistent southerly winds cannot occur everywhere in all longitudes, and neither can northerly winds. The behaviour of the atmosphere is such as to require compensating northerly winds in adjacent areas to those experiencing southerly winds, and vice versa. So that, as far as the Northwest Passage is concerned, light (or heavy) ice conditions off eastern Canada will be accompanied by heavy (or light) conditions farther west within the Archipelago. Similarly for the Northern Sea Route, light (or heavy) conditions west of Severnaya Zemlya (100°E) will co-exist with heavy (or light) conditions farther east. Nevertheless, ice-breaker escorted convoys normally ply along the western and eastern sections of the Northern Sea Route each summer and several successful passages have been made by small and large vessels through the Canadian Arctic.

However, any yacht or small vessel planning an attempt on either route should be especially reinforced to withstand moderate or light ice conditions (4/10 or less) and should be capable of hauling herself out in more severe conditions. It is clear that any such attempt should be self-supporting for several successive years since the season is short - a matter of four or five weeks. At least as far as the Northwest Passage is concerned, a radio link with Ice Control at Halifax to receive up-to-date information on the best route through the islands is essential. Even then, any attempt will still be fraught with considerable danger.

Southern Hemisphere

In late winter (September/October) pack ice extends far northwards from the Antarctic coastline to cover a vast area of the Southern Ocean. (This area is about 50 per cent greater than the sea ice area of the Arctic.) Diagram 81 shows that the winter 4/10 mean limit lies farthest north (54°S) in the longitudes of the central Atlantic and (56°S) in the central Indian Ocean. In general it lies farther south in the Pacific sector than elsewhere, possibly because the Antarctic coast also lies farther south in this sector.

As summer advances considerable melting occurs so that by February/March almost 85 per cent of the winter area will have melted. The mean summer 4/10 limit is also shown. It can be seen that there are two large areas of ice which normally persist through the summer season. These lie on either side of the Antarctic Peninsula (65°W) and are due to onshore winds in these areas, whereas elsewhere the ice is eventually removed northwards into warmer seas where it melts.

The limits shown in Diagrams 79 and 81 are taken almost completely from satellite pictures during the period 1966 to 1974. Before this time there was insufficient data to define an ice edge around the hemisphere. Many inaccurate statements have been made about the year-to-year growth in the area of Arctic ice. To a large extent these have been based on the year-to-year changes in one sector, such as the Greenland-Iceland sector, extended around the whole hemisphere without regard to the fact that compensatory conditions existed elsewhere. A recent investigation based entirely on satellite pictures, which give accurate results on a hemispherical scale, has shown that in the Arctic the area of pack ice, meaned over two consecutive three-year periods, has decreased slightly from the 1969-71 mean to the 1972-4 mean in both summer and winter.

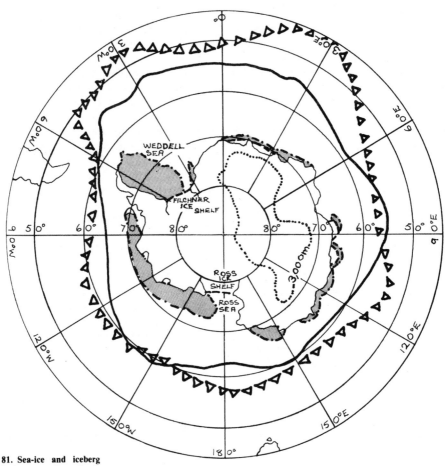

81. Sea-ice and iceberg limits – southern hemisphere

△△△△ Limit of average iceberg, spacing of 45 miles

——— mean limit of 5/10 pack ice at time of greatest extent (Sept/Oct)

—·—·— Mean limit of 5/10 pack ice at time of least extent (Feb/Mar)

········· Edge of land ice (ice front)

(There was insufficient data for an Antarctic investigation at the same time.) No positive long-period trends on the changes in polar ice can be determined until a lot more satellite data have been accumulated.

Icebergs

These large masses of floating ice are derived from floating glacier tongues or from ice shelves. The specific gravity of iceberg ice varies with the amount of imprisoned air and the mean value has not been exactly determined, but it is assumed to be about 0.900 as compared with 0.916 for pure fresh water ice, i.e. approximately nine-tenths of the *volume* of an iceberg is submerged. The *depth* of a berg under water, compared with its height above the water, varies from 5:1 for a large block-shape iceberg to 1:1 for an iceberg's remnants.

Icebergs diminish in size in three different ways; by calving, melting or erosion. A berg calves when a piece breaks off; this disturbs its equilibrium, so that it may float at a different angle or capsize. In cold water melting takes place mainly on the waterline. In warm water a berg melts mainly from below and calves frequently. Erosion is caused by wind and rain. Small pieces a few metres across are called 'growlers'; larger pieces tens of metres across are called 'bergy bits'.

In the Arctic, bergs originate mainly in the glaciers of the Greenland icecap which contain approximately 90 per cent of the land ice of the northern hemisphere. Large numbers of bergs produced from the east coast glaciers, particularly in the region of Scoresby Sound, are carried south in the East Greenland Current. Most of those surviving this journey drift round Cape Farewell and melt in the Davis Strait, but some follow south or southeast tracks from Cape Farewell particularly in the winter half of the year so that the maximum limit of icebergs occurring in April in this region lies over 400 miles southeast of Cape Farewell. However, a much larger crop of icebergs is derived from the glaciers which come down to Baffin Bay. It has been estimated that more than 40,000 bergs may be present in Baffin Bay at any one time; by far the greatest number are close in to the Greenland coast between Disko Bight and Melville Bight where most of the major parent glaciers are situated. Some of this vast number of bergs become grounded in the vicinity of their birthplace where they slowly decay; others drift out into the open summer waters of Baffin Bay and steadily decay there; but each year a significant proportion is carried by the predominant current pattern in an anticlockwise direction around the head of Baffin Bay. Of these, some ground in Melville Bight and along the east coast of Baffin Island and there slowly decay. The remainder slowly drift south with the Baffin Land and Labrador Currents, their numbers continually decreasing by grounding, or, in summer, melting in the open sea

(see Diagram 82). The number of icebergs passing south of the 48th parallel in the vicinity of the Grand Banks of Newfoundland varies considerably from year to year. Between 1946 and 1970 the number of icebergs sighted south of 48°N in that area varied from 1 in 1958 to 931 in 1957, and averaged 213 per year: the greatest number were usually sighted in April, May and June; none were sighted between September and January.

Since about nine-tenths of the volume of an iceberg is submerged, it follows that its movement is chiefly controlled by the water movement or current. However, strong winds may exert a considerable influence, by direct action on the berg and through their effect on the current. Off Newfoundland the chief factor governing the severity of any iceberg season (March to July) is the frequency of north-northwesterly winds along the Baffin Island and Labrador coasts during the months immediately preceding and early in the season. For example, in 1972 northwesterly winds prevailed from late December to May, resulting in 1587 icebergs drifting south of the 48th parallel on the Grand Banks of Newfoundland. In 1966, however, strong northeasterly onshore winds occurred in January, February and March, resulting in the grounding and eventual decay on the Labrador Coast of the 1966 iceberg crop.

In the Arctic, the irregular glacier berg of varying shape constitutes the largest class. The height of this berg varies greatly and frequently reaches 70 m; occasionally this is exceeded and one of 167 m has been measured. These figures refer to the height soon after calving, but the height quickly decreases. The largest berg so far measured south of Newfoundland was 80 m high, and the longest 517 m. Glacier bergs exceeding 1 km have been seen farther north.

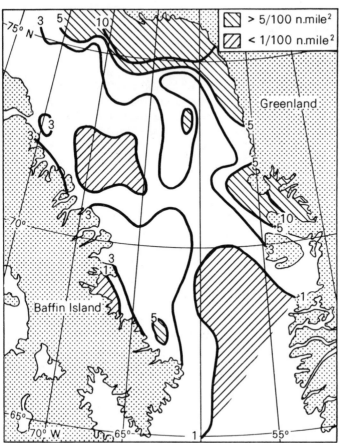

> 5/100 n.mile²

< 1/100 n.mile²

82. Iceberg census in Baffin Bay

Greenland

Baffin Island

The colour of Arctic bergs is an opaque flat white, with soft hues of green or blue. Many show veins of soil or debris; others have yellowish or brown stains in places, due probably to diatoms. Much air is imprisoned in ice in the form of bubbles permeating its whole structure. The white appearance is caused by surface weathering to a depth of 5-50 cm or more, and also to the effect of the sun's rays which release innumerable air bubbles.

'Ice island' is a name popularly used to describe a rare form of tabular berg found in the Arctic. Ice islands originate by breaking off from ice shelves, which are found principally in North Ellesmere Island and North Greenland. They stand about 5m out of the water, have a total thickness of about 30 to 50 m and may exceed 150

square miles in area; in contrast, the tabular bergs of the Antarctic commonly stand about 30 m out of the water and have a total thickness of about 200 m. Ice islands are much thicker than sea ice and those which remain in the Arctic region regain the summer loss each winter. Those which have left the Arctic basin (via the Greenland Current) have decayed by the vicinity of Cape Farewell.

The larger ice islands have hitherto been found only in the Arctic Ocean where they drift with the sea ice at an average rate of 1 to 3 miles per day. The best known, named T3 or Fletcher's Ice Island, was sighted in 1947 and has been occupied by United States scientific parties on several occasions for periods of up to two years. Since it was first discovered, and probably for many years previously, T3 has been drifting in a clockwise direction in the Beaufort Sea current system.

The 'maximum' limit of icebergs, defined as the limit beyond which icebergs are sighted at only very rare intervals, is shown in Diagram 79. On very rare occasions, icebergs or their remains may be sighted as far from the Newfoundland area as the vicinities of Bermuda, the Azores and the British Isles.

The breaking away of ice from the Antarctic continent takes place on a scale quite unknown in any other part of the world, so that vast numbers of bergs are found in the adjacent waters. Bergs here are formed by the calving of masses of ice from ice shelves or tongues, from a glacier face, or from accumulations of ice on land near the coast, fed by the flow from two or more glaciers.

Antarctic bergs are of several distinctive forms. The following descriptions should be regarded as covering only those terms which are likely to be of use to mariners. No adequate study of Antarctic bergs has yet been made.

Tabular bergs are the most common form of iceberg seen in the Antarctic. They are largely, but not all, derived from ice-shelves and show a characteristic horizontal banding. Tabular bergs are flat-topped and rectangular in shape, with a peculiar white colour and lustre as if formed of plaster of paris, due to their relatively large air content. They may be of great size, larger than any other type of berg found in either polar region. Such bergs, exceeding a mile in length, occur in hundreds. Many have been measured up to 20 or 30 miles in length, and bergs of more than twice this length have been reported. The largest berg authentically reported is one about 90 miles long, observed by the whaler *Odd I* on Januuary 7, 1927

about 50 miles northeast of Clarence Island in the South Shetland Islands. This great tabular berg was about 35 m high.

The number of bergs set free varies in different years or periods of years. There appears to have been an unusual break-up of ice-shelf in the Weddell Sea region during the years 1927-33, when the number and size of the tabular bergs in that region was exceptional. The giant berg described above was one of these. Heights up to 50 or 60 m were measured during this period.

After the southern winter of 1967 it was noticed from satellite pictures that a large section of the Trolltunga ice tongue (69½°S, 01°W) had disappeared though a giant iceberg of about the same shape was visible drifting southwest along the coast into the Weddell Sea. From 1969 to 1974 it remained grounded in about 77°S and later slowly worked its way north along the east coast of the Antarctic Peninsula, finally breaking free from the pack ice early in 1977. Since then it has moved northeast and later east. Its original size was estimated to be about 50 miles by 25 miles; on its last sighting (by the *John Biscoe)* in April 1978 it measured 38 by 16 miles, having averaged just less than ½ knot towards the northeast in the previous two weeks. The remnants were estimated to have melted in about 50°S, 0°W in December 1978, eleven years after it was calved. The Trolltunga iceberg is reckoned to be the largest ever recorded north of 55°S.

Antarctic icebergs are most frequent close to the coast; their numbers gradually decrease with increasing distance northwards. The limit shown in Diagram 81 is that for an average spacing of 45 miles between icebergs. At this limit considerable melting occurs so that few bergs are observed beyond it; the limit shown may therefore be considered a 'maximum' limit. On rare occasions icebergs have been sighted much further north, e.g. 26°S in the South Atlantic, close to the Cape of Good Hope and close to the south coasts of Australia and New Zealand.

Icebergs are sometimes extremely difficult to see, especially in moonlight or hazy conditions and of course in poor visibility. The following signs are useful at night or in poor visibility and should not be disregarded:

In the open sea the absence of developed waves in a fresh breeze may indicate that an iceberg lies to windward.

When bits of ice calve from an iceberg there is a thunderous roar.

The presence of growlers or small pieces of ice may mean that an iceberg is in the vicinity, probably to windward.

The sound of waves breaking against an iceberg may be heard on a quiet night.

Ice Accumulation

In connection with high latitude sailing there is one other form of ice, that is ice accumulating on the vessel's rigging, superstructure, deck and topsides. The rate of icing depends on the air and sea temperatures and the wind speed (sea state). Ice accumulates when water droplets, from fog, drizzle or rain, and salt water from spray or waves comes into contact with rigging, deck etc when temperature is below 0°C. Then the droplets (fresh water only) will freeze; when temperatures on the deck, rigging etc are below minus 2°C then salt water will also freeze. Since the most severe rates of accumulation are due to spray and waves a fundamental requirement for severe icing is that the air temperature and therefore that of rigging, super-structure etc, should be minus 2°C or below. The sea surface temperature should also be close to freezing (minus 2°C), but icing can still occur with sea temperatures around 10°C if air and deck temperatures are cold, say minus 10°C. The other requirement is that the winds should be strong enough to produce waves which may break over the bow or cause spray. For a yacht beating to windward the lower limit will be Beaufort Force 4. Under these conditions with air temperature minus 7°C (not impossible even in high summer) and sea temperature plus 2°C, the rate of accumulation on rigging etc will be about 3 cm in 12 hours. With stronger winds and colder temperatures the rate will be much greater, perhaps more than twelve times as much, but this is more likely in winter when no yacht should be anywhere near these regions although it is a hazard for fishing vessels.

Weather Around the British Isles, and the Fastnet Storm

THE most striking characteristic of the weather in these waters is its changeability, not only from day to day but also from one season to the same season in the following year(s). This changeability is due to the fact that the British Isles surrounding seas lie in the 'disturbed westerly' wind belt of the northern hemisphere and are therefore affected by depressions and anticyclones which move across the area with different intensities, frequencies and speeds. It is also due to their location between the Atlantic Ocean and the continent of Europe.

The day-to-day changes are sometimes very dramatic; the light winds and good weather of a settled anticyclonic period may change to gales and poor weather within twenty-four hours (especially if a yacht is sailing towards the change) while, in winter, severe depressions may follow each other in a matter of hours.

The changeability from year to year in the same season is perhaps best indicated by the weather for the Fastnet Races of 1977 and 1979. In the former race the winds were light while in the latter storm force winds swept across the Celtic Sea causing the abandonment of 24 yachts (5 of which sank) and the deaths of 15 yachtsmen. Three hundred and three yachts started the race; 85 finished. The weather events of the 1979 Fastnet storm will be described later in this chapter.

Radio Forecasts

As related in Chapter 14, changes in the weather pattern in one area, such as northwest Europe, often result from changes in another, especially upwind, region such as North America. On many occasions these changes occur rapidly and their approach will not be apparent at sea until a few hours before the change (perhaps dramatic) arrives; this may well be insufficient time for small vessels or a yacht with inexperienced crew to run for shelter. With fast-moving weather changes the yachtsman's 'horizon' is decidedly limited; it is only a few miles or a few hours.

The Bracknell shipping forecaster's 'horizon' lies thousands of miles away. He is able to study the weather developments over a vast section of the northern hemisphere and therefore to predict dramatic and often dangerous deteriorations in the weather long before there is any sign of such a change to these on board, either from the sky or the barometer.

All this adds up to the fact that there is no substitute for taking down each shipping forecast and in a rapidly developing situation, for monitoring the gale warnings transmitted at the

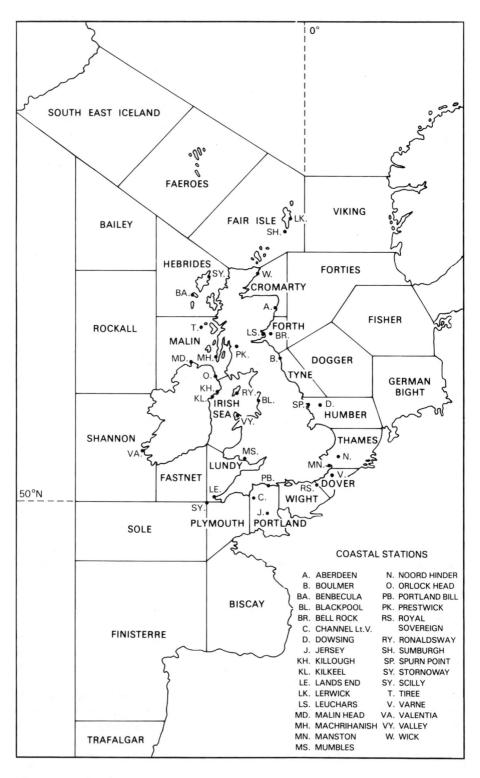

COASTAL STATIONS

A. ABERDEEN	N. NOORD HINDER
B. BOULMER	O. ORLOCK HEAD
BA. BENBECULA	PB. PORTLAND BILL
BL. BLACKPOOL	PK. PRESTWICK
BR. BELL ROCK	RS. ROYAL
C. CHANNEL Lt.V.	SOVEREIGN
D. DOWSING	RY. RONALDSWAY
J. JERSEY	SH. SUMBURGH
KH. KILLOUGH	SP. SPURN POINT
KL. KILKEEL	SY. STORNOWAY
LE. LANDS END	SY. SCILLY
LK. LERWICK	T. TIREE
LS. LEUCHARS	V. VARNE
MD. MALIN HEAD	VA. VALENTIA
MH. MACHRIHANISH	VY. VALLEY
MN. MANSTON	W. WICK
MS. MUMBLES	

North Scotland	Aberdeen Airport
East Scotland	Pitreavie, Dunfermline
Northeast England	Newcastle Weather Centre
East England	RAF Bawtry, Yorkshire *or*
	RAF Honnington, Suffolk
Midlands	Nottingham Weather Centre
Southeast England	London Weather Centre
South England	Southampton Weather Centre
Southwest England	Plymouth (Mount Batten)
West England and South Wales	Bristol Weather Centre
North Wales and Northwest England	Manchester Weather Centre
West Scotland	Glasgow Weather Centre
Northern Ireland	Belfast (Aldergrove) Airport

Table 8: Regional forecast offices around the UK

end of each hour. To continually ignore shipping forecasts will eventually put the vessel and her crew at unjustifiably high risk. Each severe storm produces a list of casualties among yachts and other craft; in nearly all these cases the storm was adequately forecast by the Meteorological Office at Bracknell. In more than a few, attention to these forecasts would have avoided disaster.

At the time of writing the UK shipping forecasts are transmitted daily on BBC Radio 4 on 200 kHz (1500 metres) four times daily. The shipping forecast areas are shown in Diagram 59 and the content of ·the forecast is discussed in Chapter 15.

Weather reports for the following coastal stations are normally given with each shipping forecast:

Tiree, Sumburgh, Bell Rock, Dowsing Lt V, Noordhinder Lt V, Varne Lt V, Royal Sovereign Lt Tr, Channel Lt V, Scilly, Valentia, Ronaldsway, Malin Head and Jersey.

Forecasts for the period until 1800 on the following day for inshore waters (up to 12 miles offshore) are also given at the end of the national programmes on Radio 4. These forecasts contain

83. Location of coastal stations (main shipping and inshore waters forecasts)

weather reports for 2200 from the following coastal stations:

England, Scotland and Wales programmes
Boulmer, Spurn Point, Manston, Portland Bill, Land's End, Mumbles, Valley and Blackpool. Prestwick, Benbecula, Stornoway, Lerwick, Wick, Aberdeen and Leuchars.

Ulster programme
Kilkeel, Malin Head, Machrihanish, Ronaldsway, Valley, Killough and Orlock Head.

In addition, for the period Good Friday to October 31 each year, warnings of winds of Force 6 or more will be transmitted by local radio stations which collectively cover the UK coastline up to 5 miles offshore. The warning, containing a time of origin, will be transmitted at the first programme break and will be repeated after the news bulletin at the end of the following hour.

Further, it is well worth while telephoning the forecaster at the nearest regional forecast office (see Table 8) before setting off from UK coasts. (Many yachtsmen who have used this free service subsequently call these offices from foreign ports as well.) The forecaster there will have all the guidance forecasts issued from Bracknell and will also have later information with which to update these forecasts as necessary; they will also have considerable knowledge of wind and weather peculiarities in their local areas.

Depression and Anticyclone Tracks

The main depression and anticyclone tracks over the eastern North Atlantic and northern Europe are shown in Diagrams 84 and 85. In all seasons the British Isles are never very far away from depression tracks, though in general they lie farther away in summer than in the other seasons.

Diagrams 86, 87 and 88 show the seasonal frequencies of vigorous,

moderate and weak depressions for the five year period 1946–50. These indicate that the more severe storms are more frequent in autumn and winter than in spring and summer, and that they are more common in northern waters than in the south. It should be noted, however, that vigorous depressions do sometimes occur in southwestern waters even in summer. The apparently higher frequencies of moderate or weak depressions in central and southern areas result from the exclusion of the more vigorous depressions which are more common in northern waters.

Somewhat surprisingly, perhaps, Diagram 85 shows that the UK lies as close to anticyclone tracks in winter as it does in summer, but what it does not show is the speed of these features. In general, anticyclones passing close to the UK in winter will move more

84. Main and secondary tracks of depressions

MAIN
SECONDARY

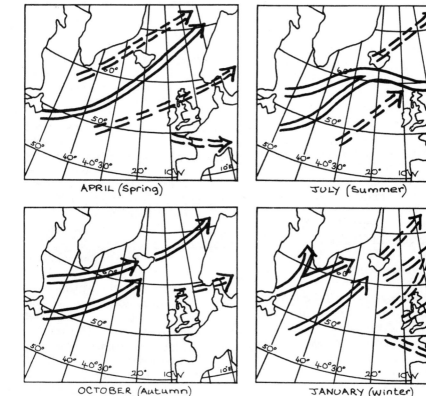

APRIL (Spring)

JULY (Summer)

OCTOBER (Autumn)

JANUARY (Winter)

quickly than in summer, so that the associated fair weather does not last as long in winter as it does in summer.

Winds

Because of the greater frequency of depressions in winter than in summer, particularly vigorous ones, it is not surprising that the winds around the British Isles are stronger in winter. It can also be seen that winds are usually stronger in the north and west than in the south and east in most seasons (Diagram 89).

The frequency of winds of Force 6 or more is shown, by 'seasonal' months, in Diagram 90 while the frequency of gales is given in Diagram 91. Both diagrams clearly indicate the higher

frequencies of strong winds and gales in northwestern waters; they also confirm that gales are more frequent in many areas in autumn and winter than they are in spring and summer, and that gales in summer in all areas are not unknown. Note that the percentage frequency of gales in July is slightly higher in sea area Fastnet than in the surrounding sea areas. This pattern also prevails in August (not shown), the Fastnet Race month.

Since most depressions pass north of the British Isles the mean direction of these winds is westerly. However, strong winds may blow from any direction on any given occasion according to the track of an individual depression. They sometimes occur on the periphery of an anticyclone or ridge of high pressure. This is particularly so in the case of a north-reaching ridge of high pressure which moves southeast

85. Main and secondary tracks of anticyclones

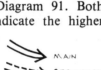

MAIN

SECONDARY

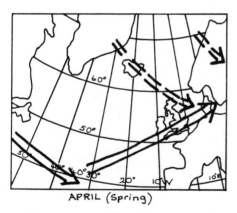

APRIL (Spring)

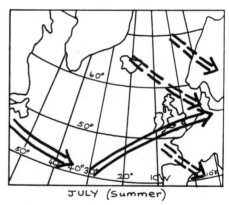

JULY (Summer)

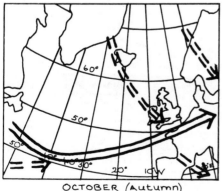

OCTOBER (Autumn)

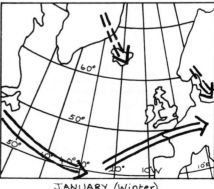

JANUARY (Winter)

across the land. This feature sometimes becomes very slow moving, or even stationary for several days, over eastern England resulting in persistently strong winds in the southern North Sea and eastern English Channel (see Diagram 92).

Visibility

The mean seasonal percentage frequency of fog (visibility less than 1 km) over these waters is shown in Diagram 93. At first sight the values may seem low, and they may well be lower than reality since many ships

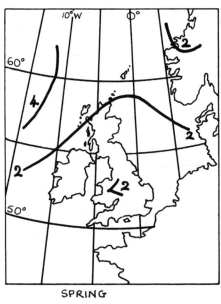

SPRING

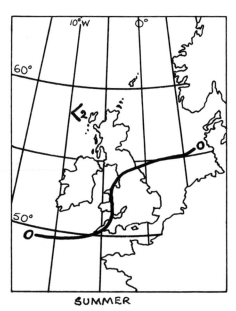

SUMMER

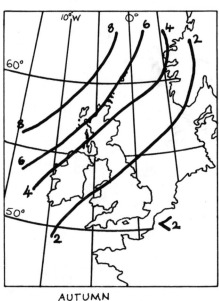

AUTUMN

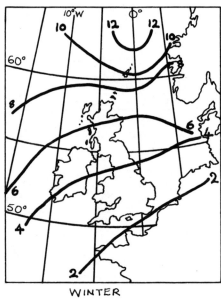

WINTER

86. Average seasonal frequency of vigorous depressions: central pressure less than 980 mb (1946–50)

which report weather regularly are too busy in poor visibility in the close approaches to UK coasts to make formal weather observations. Nevertheless the pattern from place to place within one season and from season to season is fairly realistically revealed.

Fog is least common in the autumn months September to November when frequencies average around 2 per cent and there is little geographical variation. In this season, at least for the first two months, sea temperatures

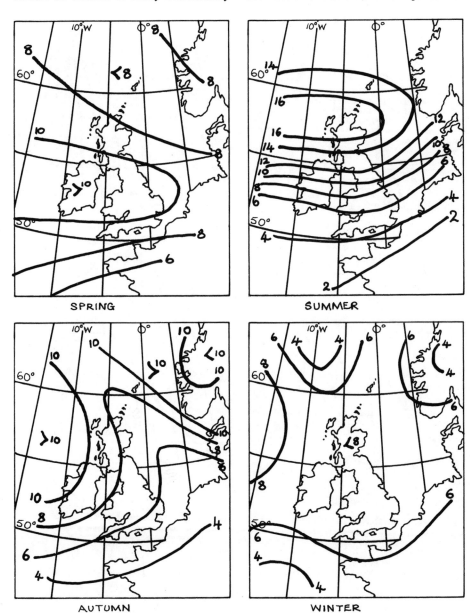

SPRING

SUMMER

AUTUMN

WINTER

87. Average seasonal frequency of moderate depressions: central pressure 981–1000 mb (1946–50)

145

are still close to their summer maximum values but air temperatures have normally begun to fall (Diagram 94). This condition, warm sea and cool air, is opposite to that required to form sea fog, which is that the sea temperature should be below the dew-point temperature of the air (which means that it must be below the dry-bulb air temperature).

As winter comes on, substantial cooling of the sea surface occurs everywhere except in southwestern, western and northern areas where the

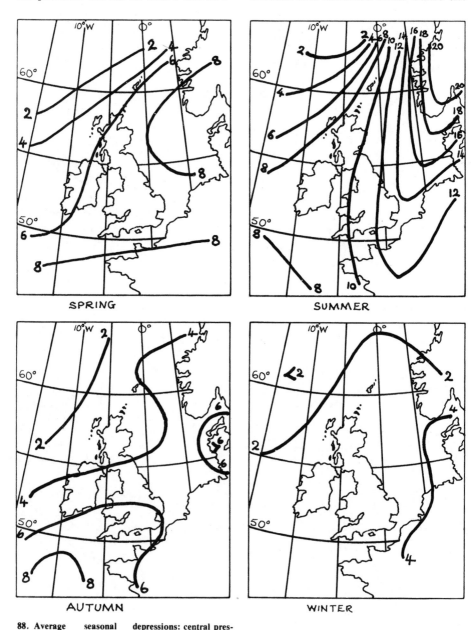

88. Average seasonal frequency of weak depressions: central pressure more than 1000 mb (1946–50)

warm North Atlantic Drift current maintains relatively high sea temperatures. The winter cooling is most marked in the central and southern North Sea. Almost any incursion of moist, mild air in winter will lead to sea fog in these areas, especially near continental coasts. Fortunately these are not prevalent as the normal winds are from some (cold) easterly point. In other areas around the British Isles the likelihood of fog is about 2 per cent, similar to the autumn value.

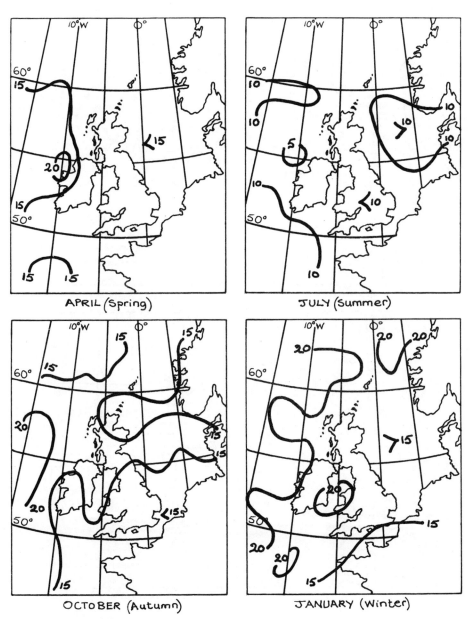

89. Mean monthly wind speeds (knots)

In spring (March to May) air temperatures recover over all areas but the sea surface temperatures lag behind staying close to their winter values. Thus in some areas mean sea surface temperatures are below the mean air temperature. This begins to favour sea fog and should the air be almost saturated (dew-point temperature almost identical to dry-bulb temperature) then sea fog may well result. This is most likely again in the eastern North Sea where the mean percentage frequency of fog is 8 per

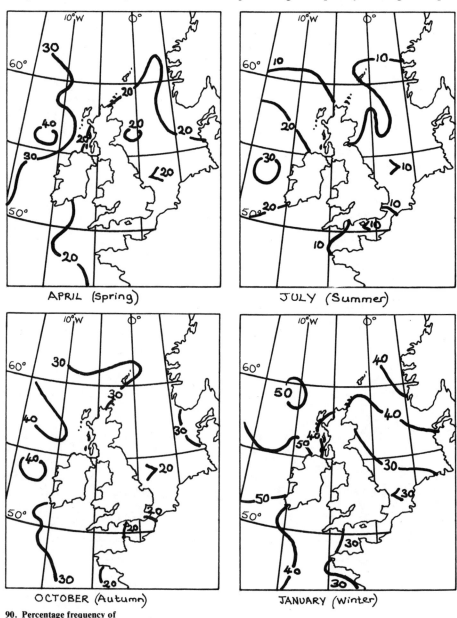

APRIL (Spring)

JULY (Summer)

OCTOBER (Autumn)

JANUARY (Winter)

90. Percentage frequency of winds of Force 6 or more

cent; in the northern North Sea the likelihood of fog now becomes greater. It should be noted that any rapid advance of tropical maritime air, i.e. strong, moist southwesterly winds, will lead to sea fog in many areas, despite the percentage frequencies shown.

In summer the percentage frequency decreases in the southern and eastern North Sea as sea surface temperatures rise towards their maximum. At this time, however, the fog risk is greatest around northern waters and, to some extent, in the western Channel and

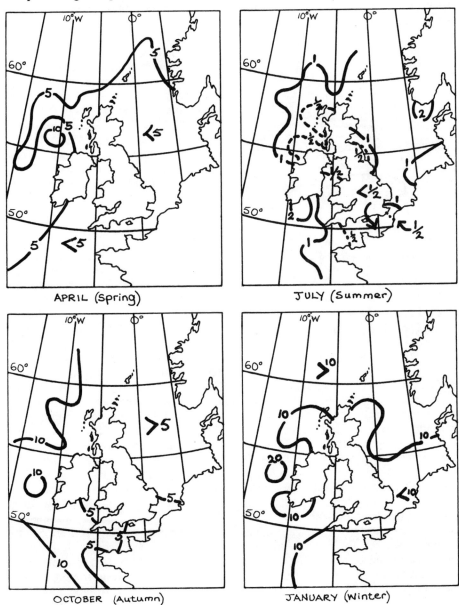

APRIL (Spring)

JULY (Summer)

OCTOBER (Autumn)

JANUARY (Winter)

91. Percentage frequency of gales of Force 8 or more

western sea areas.

By contrast land fog (radiation fog) is more common in autumn and winter. It is not uncommon, especially when land fog is widespread, for offshore winds to carry it over coastal sea areas and sometimes even farther offshore in areas such as the Dover Straits and its approaches.

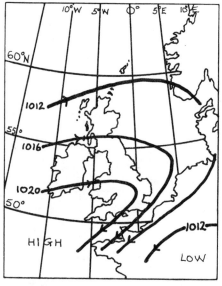

92. Synoptic situation associated with persistently strong north-easterly winds in the southern North Sea and eastern Channel

Air and Sea Surface Temperatures

To amplify the foregoing section and to indicate the mean temperature conditions for planning purposes, air and sea temperature monthly mean values are displayed by seasons in Diagrams 94a and 94b. As a point of interest it is well worth noting that at least in many southern sea areas and more particularly for the months of September and October, sea surface temperatures remain at their summer peak and therefore conditions at sea in these areas are normally as warm as they are in July and August.

Cloudiness

Similarly, for planning purposes, some mention of average cloudiness is useful. In the following paragraph 'cloudy' means a complete or an almost complete cloud cover whereas 'clear sky' means little or no cloud.

In spring and summer there are clear skies, on average, some 20 per cent of the time and cloudy skies 40-50 per cent of the time. In autumn conditions are similar in southern sea areas but clear sky frequencies are reduced to about 10 per cent in northern sea areas and as a consequence cloudy days increase to a little over 50 per cent. In winter the weather is cloudy for 50-60 per cent of the time and clear for about 15 per cent in all sea areas. It is a common experience that persistently strong winds on a cloudy day seem to be one Beaufort force less when the sun comes out!

Currents

The flow of water around the British Isles is chiefly due to tidal influences. (For information on streams and ranges the reader is referred to the Admiralty Tide tables and to their various Tidal Stream Atlases.) Nevertheless, there is a weak counter-clockwise circulation around the North Sea of less than ¼ knot, and there is a net flow of water eastwards along the Channel and through the Dover Straits at an even slower rate. But these are only of academic interest to mariners.

It can be seen, then, that sailing conditions are generally quite favourable in these waters. The winds average Beaufort Force 3 to 4, it is relatively warm, at least in summer and early autumn, and fairly sunny. Even in winter, given adequate clothing and cabin heating, sailing conditions are generally far from impossible. Nevertheless, since severe depressions may suddenly arrive in these waters in winter, it is prudent to have numerous bolt-holes in mind where one can run for shelter at any stage of a cruise.

Sudden storms, however, are not

confined to the winter season; they may also occur in summer as the 1979 Fastnet Race has so disastrously indicated. We shall pause here to consider the behaviour of this unusually severe mid-August depression.

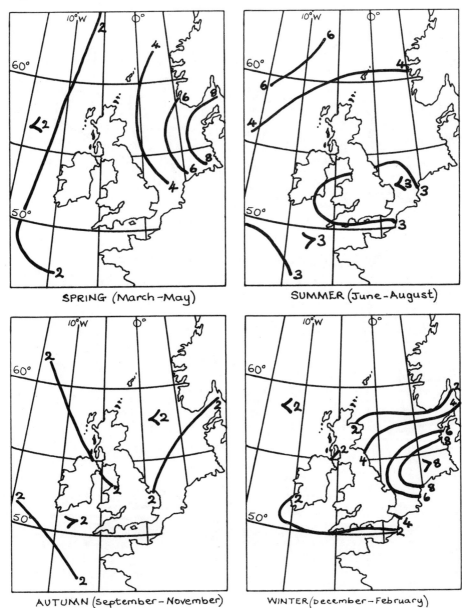

SPRING (March–May)

SUMMER (June–August)

AUTUMN (September–November)

WINTER (December–February)

93. Average seasonal percentage frequency of fog

The Fastnet Storm

During the 1979 Fastnet Race, on the night of August 13-14, when the fleet was spread out across the area between Land's End and Fastnet Rock, a vigorous Atlantic depression (centre 976 mb) crossed sea area Shannon and Southern Ireland. The accompanying storm force winds of late Monday (13th) evening to early afternoon of Tuesday (14th) wreaked havoc throughout the fleet. From the 303 yachts that started only 85 finished the race. From the remainder, 24 yachts

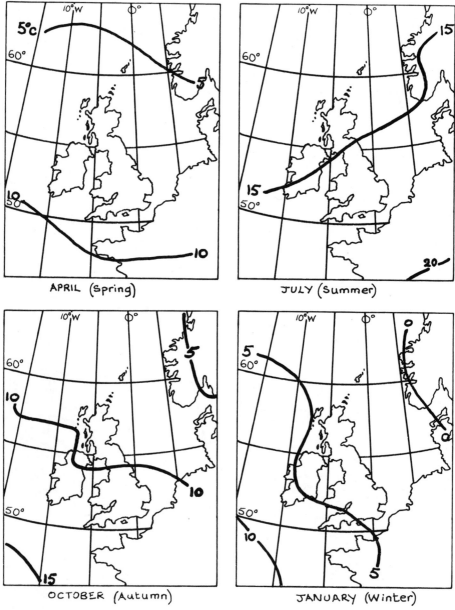

APRIL (Spring)

JULY (Summer)

OCTOBER (Autumn)

JANUARY (Winter)

94a. Mean monthly air temperature (°C)

were abandoned (5 subsequently sank) and 15 yachtsmen lost their lives.

The track of the depression is shown in Diagram 95a, while the synoptic situation at 0400 BST on the 14th is in Diagram 95b. The Plates include a polar-orbiting satellite's view of the storm on the previous afternoon. All times in this discussion are local summer time, i.e. GMT + 1 hr.

At the time of the start of the race at 1330 on Saturday August 11 the depression was located over Nova Scotia, having developed a few days

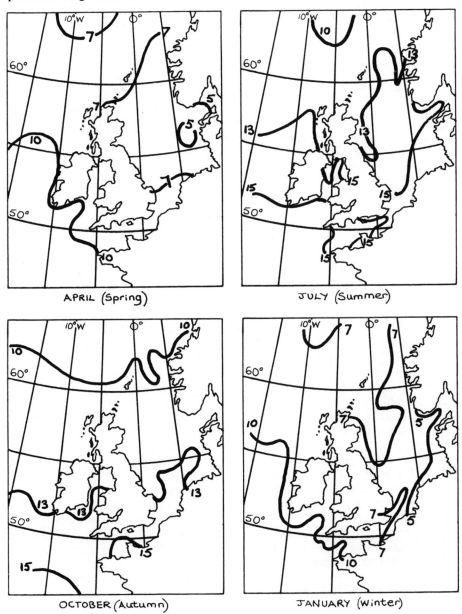

APRIL (Spring)

JULY (Summer)

OCTOBER (Autumn)

JANUARY (Winter)

94b. Mean monthly sea surface temperature (°C)

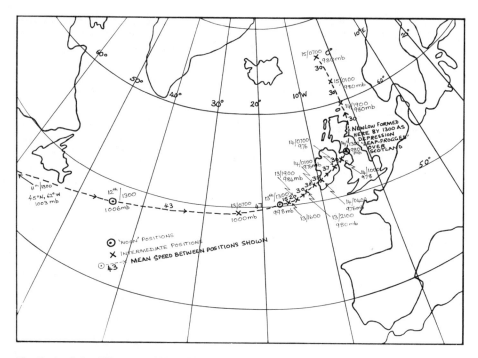

95a. Track of the 1979 'Fastnet' depression, noon 11th to 0700 on 15 Aug 1979

earlier in the lee of the Rocky Mountains. It was already moving at about 35 knots on a mainly easterly course.

When the leading yachts had reached the Lizard 24 hours later (midday on 12th) the depression was centred about 300 miles east of Newfoundland. It received its first mention in the Radio 4 shipping forecast at 1750 which stated that the low was expected to move rapidly (35 to 45 knots) to a position 350 miles southwest of Valentia by 1300 on Monday 13th and that it would deepen, very slightly, to 998 mb. (A depression moving rapidly towards sea areas Shannon and Sole was clearly of vital interest to the race fleet.)

The first visible signs of the approaching depression became apparent around daybreak on Monday 13th when the wind had backed to southwest and the barometer had begun to fall. (The approaching upper cloud would not have been visible due to the cloud associated with a previous frontal system which was still crossing the area - see Plates.) At this time most of the fleet lay to the westward of Land's End and the low was located about 600 miles west-southwest of Valentia. The forecast at 0625 that day stated that the depression was expected to move rather quickly (25 to 35 knots) northeast to sea area Rockall, central pressure 1000 mb, by 0100 Tuesday 14th. Winds of Force 5 to 6 were forecast for the relevant sea areas. The 1355 forecast that day (Monday 13th) gave similar information. During this six-hour interval the depression (998 mb) had moved east-northeast at over 45 knots to a position about 300 miles west-southwest of Valentia.

However, even as the 1355 forecast was being broadcast by the BBC, southwesterly Force 8 gale warnings were being issued from Bracknell for sea areas Sole, Fastnet and Shannon (and later for Irish Sea and Lundy) for

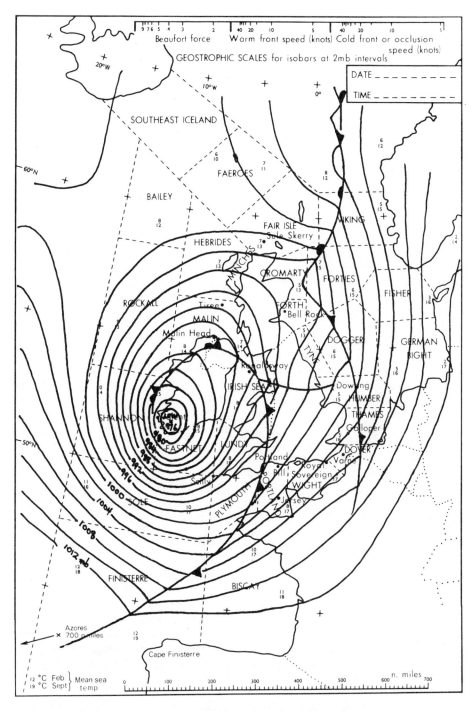

95b. The 'Fastnet Storm' at its height at 0400 BST on 14 Aug 1979. Wind speeds may be estimated by measuring distance between isobars with dividers and then 'measuring off' using Beaufort scale at top of page. For curvature correction see Chapter 4.

by then it had become clear that the depression, quite unexpectedly, had abandoned its mad gallop across the Atlantic (at over 40 knots for most of the time) and had now slowed down to between 10 and 20 knots. It was also deepening.

During the following 12 hours (midday 13th to midnight 14th) its central pressure dropped from 998 mb to 976 mb and the low had moved to a position about 30 miles west of Valentia. This deepening, by any standards, is an unusual event in the eastern North Atlantic in summer. The significance of this event is perhaps best understood by comparing Diagram 95b (the actual situation at 0400 on the 14th - the height of the storm) with 95c (a hypothetical situation in which there had been no deepening and in which the positions of the fronts have accordingly been adjusted). The stronger gradients in the reality are strikingly apparent. The more violent winds and seas were certainly more apparent to the racing fleet. During the evening of the 13th the depression's cold front swept across the race area but due to the fact that there was little troughing of the isobars along this particular front there was little or no veer of wind as the front passed through.

But to return to the afternoon and evening of the 13th, gale warnings were issued at 1355 from Bracknell, though communications delays at this time prevented them from being broadcast till 1505. However, this warning still gave several hours' notice that the Force 6 to 7 winds then being experienced would increase to gale force, though by the time it was being broadcast the pressure falls in the area must have indicated that this was inevitable. During the afternoon the leading yachts were over halfway between Land's End and Fastnet Rock and at 1330 *Innovation* logged the arrival of Force 7 winds. After a slow first two days, most of the fleet was now romping along in freshening southwesterly winds.

At 1750 the southwesterly gale warnings, now expected to veer northwest later, were repeated for the race sea areas. But at around this time the full extent of the deepening had begun to be revealed and so at 1805 severe gale Force 9 warnings were issued. At 1900 the low (984 mb) was centred about 200 miles from Valentia and at 2000 *Oystercatcher* logged the onset of Force 9 winds.

During the early evening the depression, still deepening, began to accelerate to about 35 knots northeast and later eastwards (Diagram 95a). As a result even stronger winds and extremely rough seas prevailed in the race area for the following 18 hours or so. Storm Force 10 warnings were issued at 2245, just ahead of the onset of 45-50 knot (Force 9 to 10) winds logged by *Oystercatcher* at 2300 when she was 50 miles south of the Rock.

But by then, due to the chaotic state of the sea, some yachts were already in difficulty. *Camargue* was knocked down for the first time at 2100 hours and after further knockdowns was later abandoned. During the next 12 hours or so many other yachts suffered severe damage, some foundered, and some lost crew members. The leading yachts rounded the Rock at about midnight and conditions for them became a little less dangerous as they raced homewards. For most yachts racing tactics gave way to survival tactics as the low passed over Valentia and crossed Southern Ireland maintaining its central pressure of 976 mb and its storm force winds, which veered from southwest to northwest in sea area Fastnet as the low moved east across Ireland.

From several reports it is clear that huge waves seemed to suddenly arrive in the race area and that these became more violent and chaotic during the early hours of Tuesday 14th. Diagram

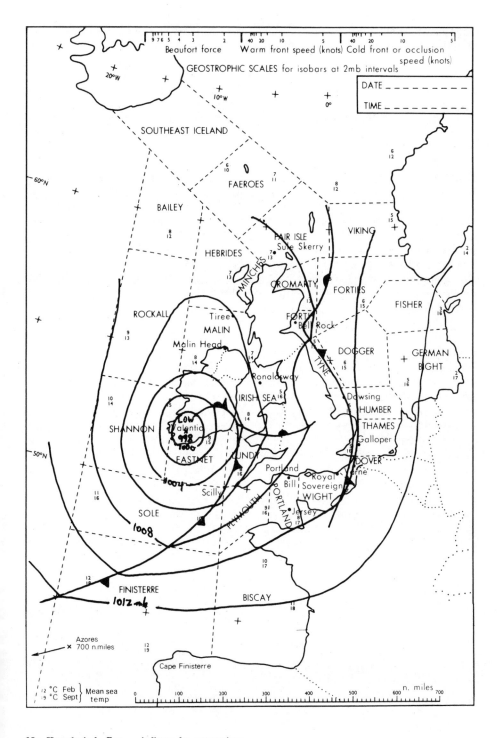

95c. Hypothetical Fast-net Storm without 'deepening stage', to indicate by comparison with 95b the severity of the real storm.

73 indicates that wave height is proportional to the duration of the wind and its fetch as well as to wind speed. Fetch is important only up to a few hundred miles; after that it becomes a minor factor.

As stated in Chapter 18, the large waves in a rapidly moving depression may well be travelling at the storm speed and are therefore continuously maintained by the very strong winds so that the arrival of these large seas occurs simultaneously with the onset of the very strong winds: in these circumstances the duration of the very strong winds in a given area is unrelated to wave height. Further, the slowing-down and deepening of the depression must have introduced several new wave trains to make the resultant sea state even more chaotic and violent. Moreover, the veer of the storm force winds from southwest to northwest during Tuesday morning must have added even more chaos to the sea surface. But there is one remaining factor: the seabed topography and its effect on the waves.

The larger waves of this storm system (moving at about 35 knots) probably had a period of about 12 seconds. The associated (Chapter 18, Table 7) wave length would probably be about 225 m. Since waves are 'unaware' of the seabed until the water depth is less than half the wavelength, the large waves of this storm would only be affected in depths of less than about 110m (55 fathoms). The depths over many of the banks in the area between Fastnet and Land's End is less than 50 fathoms. The larger waves would be slowed down over these banks, their leading edges would thus become steeper until the waves would be unstable and their upper portions would break. At 0200 on the 14th *Griffin* was capsized by a huge wave, the upper 3 to 4 m of which was observed to be breaking. It is probable that most yachts observed, if not experienced, such breaking waves.

During the remainder of Tuesday 14th the depression continued northeast at about 30 knots across the Irish Sea and over southern Scotland and later turned north to run past the Shetlands at about 2000.

The Fastnet Storm itself is not unique. Such vigorous depressions may be expected to affect the Southwest Approaches or the English Channel in summer once in every five to ten years or so. The chaotic and huge seas associated with such storms will become particularly dangerous to small vessels especially in areas where the water depth is less than half the wavelength of the larger waves.

CHAPTER 22

Weather in the Baltic

FOR the purposes of this book the whole area from eastern Denmark to the Bay of Bothnia will be called the Baltic: the proper names for the various parts of this area are shown in Diagram 96.

The climate of the Baltic, though it too lies in the 'disturbed westerlies' wind belt, is more extreme than in the British Isles. Due to its location on the edge of an ocean the latter experiences a maritime climate - mild winters, cool summers, rain more or less evenly distributed throughout the year, and fairly windy - whereas the continental climate of the Baltic is typified by cold winters (sea ice forms over northern areas in most winters), warmer summers and less wind in summer than in the UK. Sailing is possible, at least in southern areas, throughout the winter, though it is then often bitterly cold and the possibility of superstructure icing should be kept in mind. In summer the sailing is normally very good although winds are often a little light. However, as will be disussed later, severe depressions do sometimes affect the area more especially early and late in the summer season.

Radio Forecasts
As always, there is no substitute for taking down shipping forecasts. The area is particularly well served by Germany (both Federal and Democratic), Sweden and Denmark; many of these transmissions are in English. Most broadcasts from Sweden are in their own language; note that their wind forecasts are invariably given in metres per second and not Beaufort forces (2 knots = 1 m/sec). Shipping forecasts are also transmitted by Russia, Poland and Finland, the latter including warnings of winds in excess of Beaufort Force 5 in the Gulfs of Bothnia and Finland and the north Baltic Sea. The only country to cover the whole area from the Skagerrak to the Bay of Bothnia is Sweden. Germany covers the region from the Skagerrak to the northern Baltic Sea (about 60°N) while Denmark covers much of the North Sea, the Skagerrak, Kattegat and the southern part of the Baltic Sea. Transmission times and frequencies, which are liable to change from time to time and are therefore not included here, are given in the Admiralty List of Radio Signals, Vol 3 and other similar publications, and less complete listings are found in *Reed's* and other nautical almanacs.

Depression and Anticyclone Tracks
Diagram 100 shows that depressions cross the Baltic from southwest to northeast in all seasons and that in summer and autumn a main

96. Sea areas of the Baltic

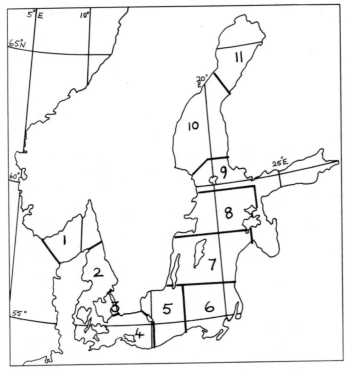

97. Shipping forecast areas – Sweden
  1 Skagerrak
  2 Kattegat
  3 The Sound
  4 Southern Baltic W of Bornholm
  5 Southern Baltic, between Bornholm and Midsjö Banks
  6 Southern Baltic, E of Midsjö Banks
  7 Central Baltic
  8 Northern Baltic
  9 Sea of Aland
 10 Sea of Bothnia
 11 Bay of Bothnia

98. Shipping forecast areas – Denmark
1 Baltic E of Bornholm
2 Baltic W of Bornholm
3 The Sound
4 Belt Sea
5 Kattegat
6 Skagerrak

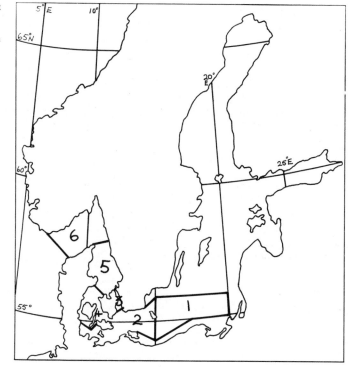

99. Shipping forecast areas – Germany (FDR and DDR)
1 Northern Baltic
2 Eastern Baltic
3 Mid-Baltic
4 Western Baltic
5 Kattegat
6 Skaggerak

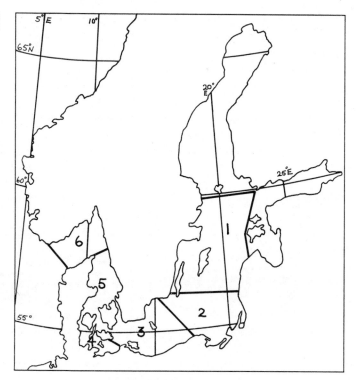

161

(preferred) track lies across the area. The seasonal frequency of vigorous, moderate and weak depressions for the period 1946-50 is given in Diagrams 102 to 104. Vigorous depressions (central pressure less than 980 mb) and their associated gale force winds are extremely rare in summer and autumn: during the five-year period 1946-50 there were none over the whole region eastwards of the Kattegat. In winter and spring vigorous storms are somewhat more frequent especially in central and northern areas.

In practically every season the frequency of moderate or weak depressions is greatest in the area of the Skagerrak and Kattegat and then decreases northeastwards farther into the Baltic. The high frequencies of weak circulations in summer are probably chiefly due to heat lows which readily form in that season.

Diagram 101 indicates a main anticyclone track (west to east) south of the Baltic with secondary tracks (northwest to southeast) in all seasons.

In general, the whole area is more likely to be affected by anticyclones in winter with depressions then making

100. Main and secondary (broken lines) tracks of depression over the North Sea and Baltic

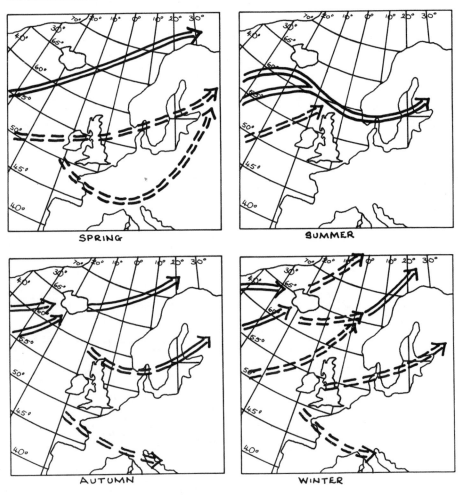

SPRING

SUMMER

AUTUMN

WINTER

occasional incursions into southern regions whereas in summer the continental high gives way and no longer blocks the advance of depressions from the west. However, severe depressions in winter occasionally overcome the blocking effect of the continental high and advance over the whole region.

Winds

Diagram 105 displays mean monthly wind speed for the months of April, July, October and January, as representative of the spring, summer,

autumn and winter seasons. It can be seen that mean winds are least in summer and steadily increase to a maximum of around 15 knots or so in winter. They are stronger in southwestern areas and weakest in more northeastern areas in all seasons.

To a very large extent this pattern is also confirmed in Diagrams 106 and 107. The frequency of strong winds of Force 6 or more can be seen to decrease from the Skagerrak into the Kattegat and from south to north within the Baltic Sea and northern gulfs. Diagram 107 (gale frequency) confirms this

101. Main and secondary (broken lines) tracks of anticyclones over the North Sea and Baltic

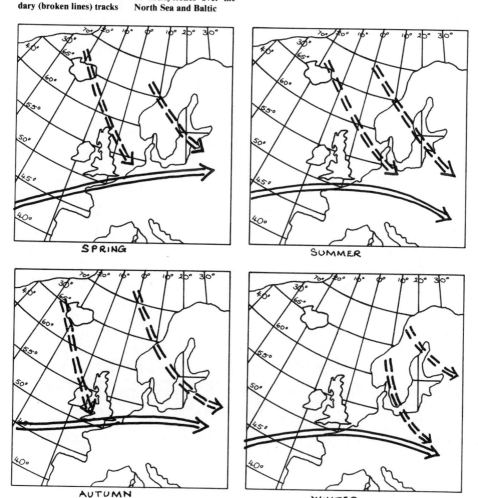

SPRING

SUMMER

AUTUMN

WINTER

pattern though in October (autumn) the frequency of gales is almost uniform over the whole region.

The directions are variable due to the numbers of depressions which affect the region and their variable tracks; nevertheless, in the long term they are predominantly from some westerly point, though in winter the winds are not infrequently from the east. It should be noted here that the winds in the Gulfs of Bothnia and Finland normally blow along the gulfs either towards their heads or in the reverse

102. Average seasonal frequency of vigorous depressions – central pressure less than 980 mb (1946–50)

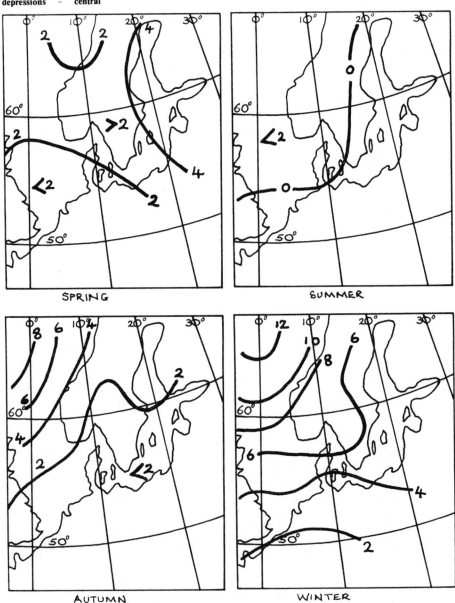

SPRING

SUMMER

AUTUMN

WINTER

direction according to the general direction of the pressure gradient. In summer the winds are generally very light - probably the only problem when cruising in the Baltic during this season. Rather than drift in central sea areas, it would often pay to stand inshore to use the sea-breezes which usually develop on otherwise light wind days.

Visibility

The average percentage frequencies of fog for April (spring), July (summer),

103. Average seasonal pressure 981–1000 mb frequency of moderate (1946–50) depressions – central

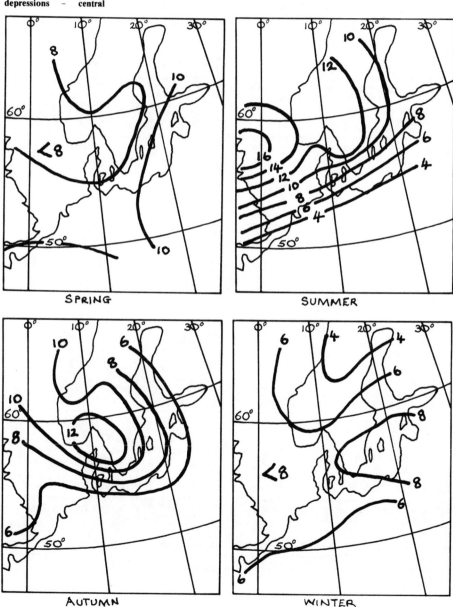

SPRING

SUMMER

AUTUMN

WINTER

165

October (autumn) and January (winter) are shown in Diagram 108. This clearly indicates that the highest frequencies occur in winter and spring. In these seasons the sea surface temperatures are low so that almost any incursion of tropical maritime air into the region will result in sea fog. These incursions are less likely in the more northern areas so the frequency of fog is less in these regions at this time.

Poor visibilities may sometimes occur downwind of an industrial

104. **Average seasonal frequency of weak depressions – central pressure over 1000 mb (1946–50)**

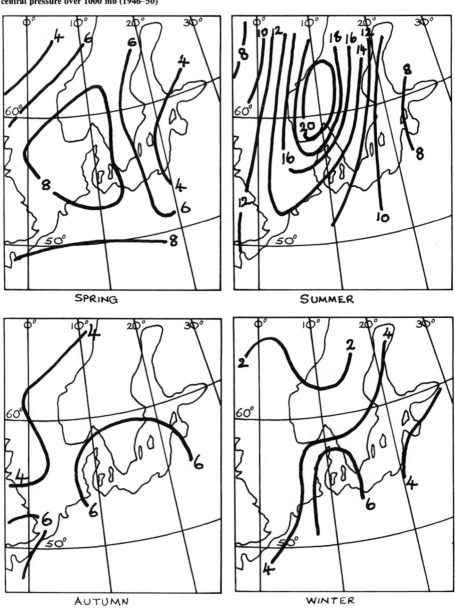

SPRING

SUMMER

AUTUMN

WINTER

region when a temperature inversion exists (see Chapter 13) and may also occur in heavy rain, especially in summer thunderstorms and of course in snow in winter.

Air and Sea Surface Temperatures
As an indicator of the 'continentality' of the climate of this region, mean air temperatures show a considerable drop from above 15°C/59°F in July to below 0°C/32°F over the whole region, and to minus 10°C/14°F in the

105. Mean monthly wind speeds (knots)

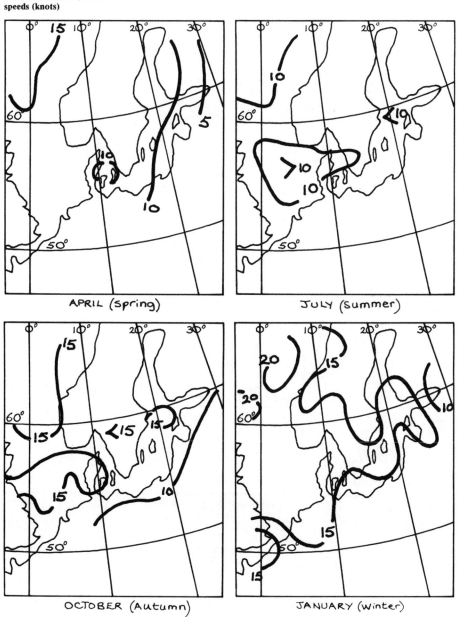

APRIL (spring)

JULY (summer)

OCTOBER (Autumn)

JANUARY (winter)

far north in January (Diagram 109a). Due to the relatively shallow water depths, sea surface temperatures also show a considerable range from summer to winter (Diagram 109b).

Sea Ice and Icing

The Baltic is affected by sea ice to a lesser or greater extent each winter. Due to the narrowness of the Belts and Sounds in the Kattegat which restrict the inflow of oceanic water of normal salinity (35ppt) into the region and to

106. Percentage frequency of winds of Force 6 or more

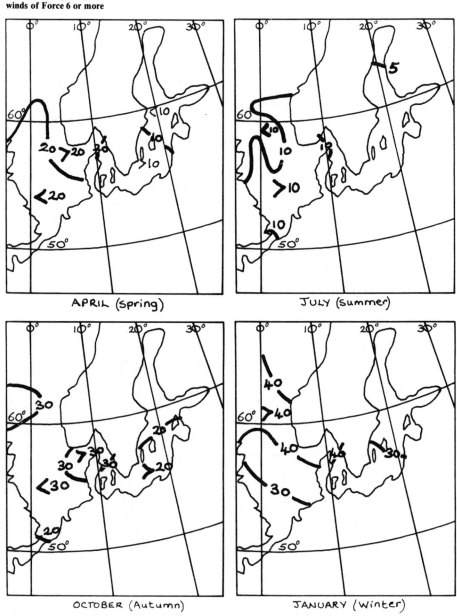

APRIL (Spring)

JULY (Summer)

OCTOBER (Autumn)

JANUARY (Winter)

the number of rivers which feed fresh water into it, the salinity of the Baltic is very low (about 5ppt over practically the whole region east of Denmark). Sea ice normally first forms at the head of the Bay of Bothnia in December. The sea ice normally reaches its greatest extent in February in the Baltic area, when it covers the whole of the Gulfs of Finland and Riga and most of the Gulf of Bothnia. At this time there is usually some thin ice in the Danish Belts and Sounds. Sea ice normally melts over the whole region during

107. Percentage frequency of gales of Force 8 or more

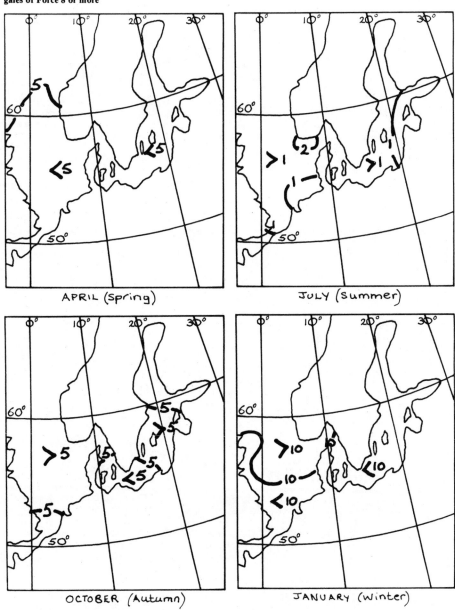

May. The normal limits of approximately half-cover of sea ice are shown in Diagram 110. In an extremely cold winter sea ice covers practically the whole Baltic region including the Kattegat and also affects the western coasts of Denmark. Any plan to winter in the Baltic region should therefore include the requirement to haul out to avoid the risk of damage by sea ice.

Sailing is possible, despite the very cold conditions, in the southern Baltic in an average winter; it should be remembered then that due to the very

108. Average monthly percentage frequency of fog

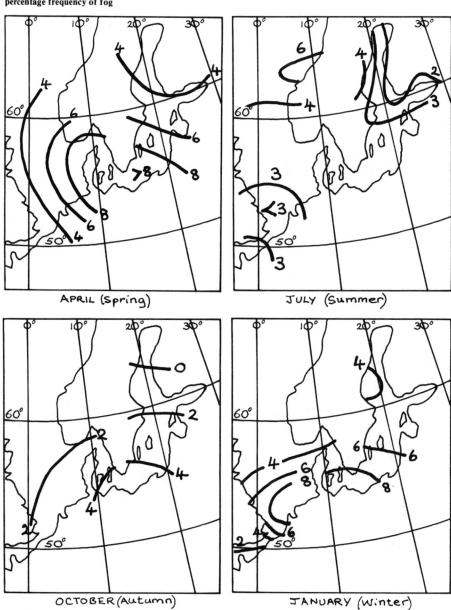

APRIL (Spring)

JULY (Summer)

OCTOBER (Autumn)

JANUARY (Winter)

low air temperatures superstructure icing may become a serious problem in all but light winds. This is particularly so when easterlies of moderate strength or greater prevail over the region. Then air temperatures will probably be well below the mean figures shown in Diagram 109a. The hull and rigging will have the same temperature as the air and so any water taken on board as spray or broken waves will freeze. The problem becomes much greater, of course, when going to windward.

109a. Mean monthly air temperatures (°C)

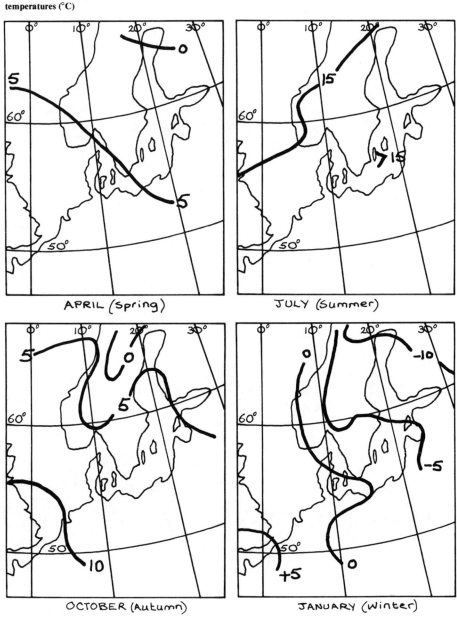

APRIL (Spring)

JULY (Summer)

OCTOBER (Autumn)

JANUARY (Winter)

Currents

The currents of the region are weak and of little interest to yachtsmen, but there is said to be a counter-clockwise circulation of water around the Baltic at a rate of less than ¼ knot. Tidal streams are also weak. However, at times there is a sudden rise or fall of up to a metre or so especially in southern areas, which is chiefly due to the effect of strong winds restricting the flow of water through the channels between Denmark and Sweden. At the cessation of the strong winds the water surges

109b. Mean monthly sea surface temperatures (°C)

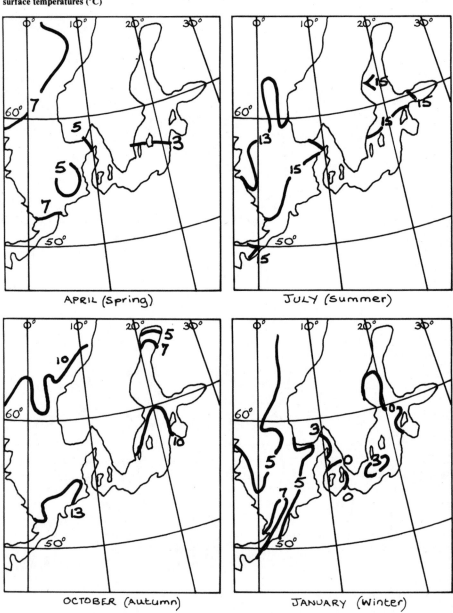

APRIL (Spring)

JULY (Summer)

OCTOBER (Autumn)

JANUARY (Winter)

110a. Average limit of 4/10 pack ice at end of December, January and February

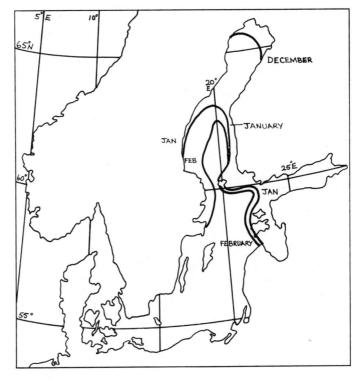

110b. Average limit of 4/10 pack ice at end of March and April

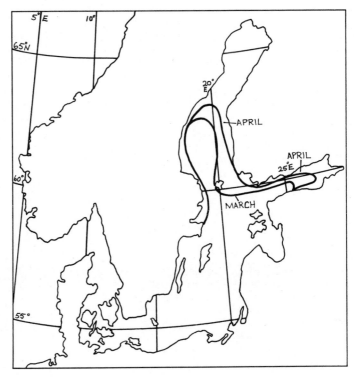

through the channels into or out of the Baltic according to the lately prevailing wind directions. For more details on this effect and on the variable and sometimes strong flow through the Belts and Sounds, the reader is referred to the appropriate Sailing Directions.

Cloudiness

As it is always more pleasant sailing in sunny weather, the following information may be useful. In spring and summer skies are almost clear for about 30 per cent of the time and almost completely cloud-covered again for the same length of time. In autumn the percentages are 20 and 50 respectively, and in winter though the percentage frequency of clear skies is again 20 per cent, cloud-covered occasions increase to about 60 per cent.

CHAPTER 23

Weather Along the Routes to the Mediterranean, the Canary Islands, Madeira and the Azores

THE enjoyment of setting off south or southwest towards the Mediterranean or the oceanic islands (Canaries, Madeira and the Azores) is increased by the fact that generally the risk of experiencing the two main weather hazards of strong winds and fog steadily decreases with increasing distance from the Southwest Approaches, that it will become warmer and that there will be more sunshine.

The climate of these routes varies from the disturbed westerlies in the north to the sub-tropical (horse latitude) high pressure belt in the area of the oceanic islands: in general, from often windy wet weather to warm sunny weather with light winds. Strong winds, as we shall later see, may sometimes affect even the southern portion of the routes. Moreover light winds or calms within anticyclones are also a bother to yachtsmen and they too are to be avoided if at all possible. This is best done by following the weather patterns and forecasts transmitted from a number of stations along and outside, the routes.

Radio forecasts

The normal British Isles shipping forecasts broadcast on BBC Radio 4 (1500m) cover the routes down to lat 35°N (about 45°N on the Azores route). The North Atlantic Weather Bulletin, prepared at Bracknell and relayed in Morse code figures by Portishead Radio, and the Fleet forecast from Whitehall, cover the remainder of the routes (Diagrams 111, 112). It is well worth listening to these oceanic forecasts for several days before setting off and during the early stage of the cruise so as to become familiar with the development and progression of highs and lows over the routes, and to practise transcribing them, which is made easier with a tape recorder. Portugal and Spain also transmit forecasts for their offshore areas and the areas around the oceanic islands. Though the broadcasts from Portugal are sometimes in English, those from Spain are always in their own language. Forecasts covering a wide area centred on the Azores are broadcast from Ponta Delgada.

On passage to the islands it may be useful to listen to weather forecasts for the area west of 35°W including the Caribbean for information on hurricanes, transmitted by the US Navy from Londonderry, Rota (Spain) and Athens, as well as several major national centres on the other side of the ocean such as Miami and Norfolk, Virginia. Full details of all these forecasts are given in the Admiralty List of Radio Signals Vol 3 and in

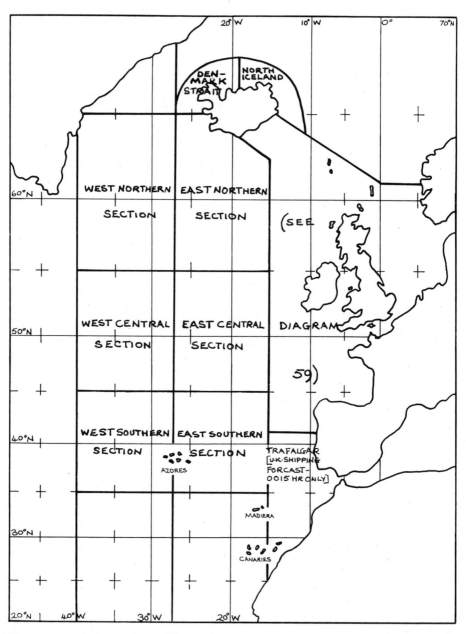

111. Forecast areas for UK North Atlantic weather bulletin

World-wide Marine Weather Broadcasts, published by NOAA in the USA.

The cost of marine radio facsimile equipment is now roughly comparable with that of satellite navigation and radar equipment. Weather facsimile machines, also called chart recorders, automatically produce analysis and prognosis (forecast) charts now being transmitted by many nations several times each day. Charts of forecast

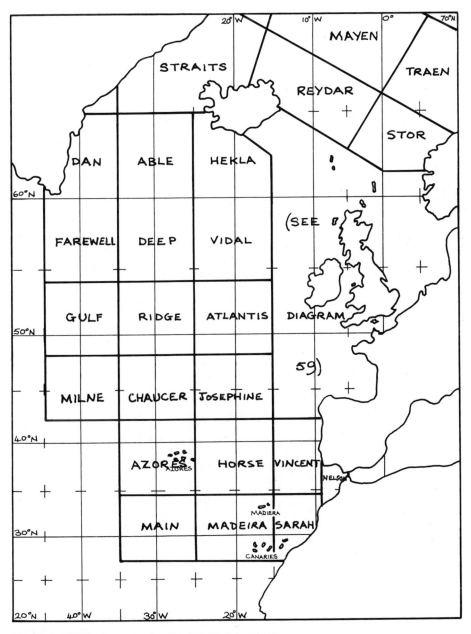

112. North Atlantic forecast areas for UK Fleet broadcasts

wave heights, sea surface temperatures, etc are also broadcast by some nations. The two lists named above give details of transmissions and content. (The North Atlantic and much of the North Pacific are collectively covered by these broadcasts.)

By listening to the forecasts covering the routes, which will include the position and movement of highs and lows, a course may be shaped to avoid the strong winds and rough seas on the

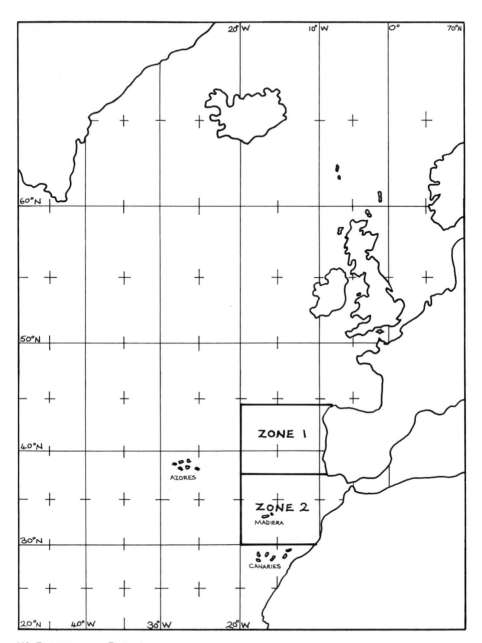

113. Forecast areas – Portugal

unfavourable side of a depression and also the light winds within an anticyclone.

In the absence of such forecasts, the guidance of Chapters 8 and 9 may be used to avoid bad weather. The earliest signs of the approach of a vigorous depression will probably come from the swell advancing ahead of the storm centre, though as there is normally some swell in the Southwest Approaches and further offshore this

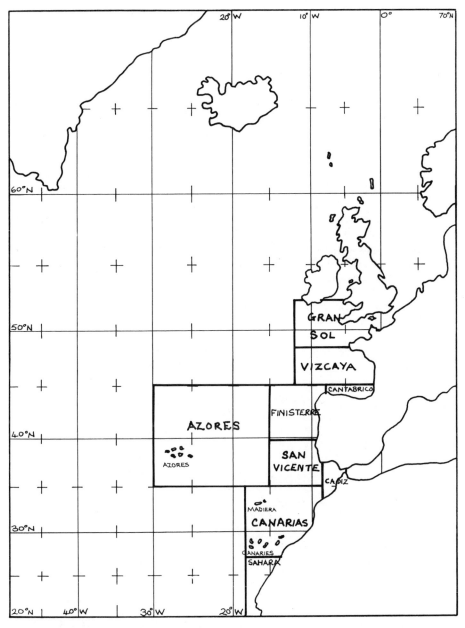

114. Forecast areas – Spain

may be difficult to detect initially. A persistent swell from one direction, especially if increasing in height, is a reasonably good indicator that a vigorous depression is approaching. That is the time for a decision on whether to take avoiding action and if so in which direction. If the depression appears to be approaching from the southwest (southwesterly rising swell) then by this time the wind will have already backed into the southwest, or

179

perhaps even further, so that the rhumb line course is no longer tenable. Obviously, then, port tack becomes the favoured tack thus allowing one to pass to the north of the depression so as to make use of the easterly winds and following seas before resuming a new course for the destination. Meantime, barometer, state of sky and wind direction should be frequently monitored to assess the relative movement of the depression.

In the case of an anticyclone in the path of a yacht, there are no indicating signs (a rising barometer is not conclusive) and, in the absence of forecast information, the yachtsman is left with course alterations which will largely be pure guesswork.

Depression and Anticyclone Tracks
As progress is made south and southwestwards, the risk of being affected by a vigorous depression steadily decreases while the chance of coming under the influence of an anticyclone steadily increases. Though no main tracks of depressions lie across the routes, a secondary track crosses the Southwest Approaches and Biscay in all seasons except perhaps in summer. Occasionally, and more especially in winter, slow-moving or stationary depressions develop on the western side of the Iberian peninsula (Diagram 116). They may remain in the area for several days or even a week or two.

Hurricanes occasionally wander into the vicinity of the Azores in the months between May and November. Typical tracks are shown in Diagrams 115. (See Chapter 17.)

Anticyclone tracks clearly favour the Azores region. In fact, the Azores high is one of the semi-permanent subtropical high pressure areas. Occasionally this high extends north and east perhaps as a ridge or as a separate high. Though the tracks

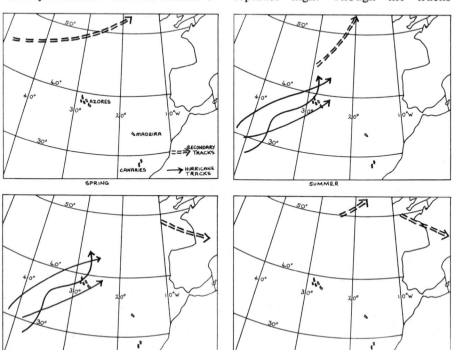

115. Tracks of depressions and examples of hurricane tracks

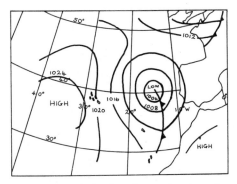

116. Typical example of a slow-moving depression in winter, off Iberia

shown in Diagram 117 suggest, perhaps, constantly moving anti-cyclones, there is normally a large anti-cyclone in residence somewhere in the region south and west of the Azores.

Pressure

The persistence of an anticyclone in the Azores area is confirmed by Diagram 118. It also indicates the mean wind directions and the change to northerly winds off Portugal especially in summer - the Portuguese Trades.

Winds

In winter the predominant wind direction is southwesterly along the route to the Azores, but changes to become light and variable south of about lat 39°N along the other routes. In spring the pattern is similar but winds tend to become more northerly along and southward of the Portuguese coast. Winds become more westerly, on average, in the north and along the route to the Azores while the Portuguese Trades become more predominant over the other routes in summer. In the north the autumn winds become more southwesterly while the Portuguese Trades become less dominant.

The Portuguese Trades are most persistent from June to August. In these months their high constancy is as much due to the formation of a heat

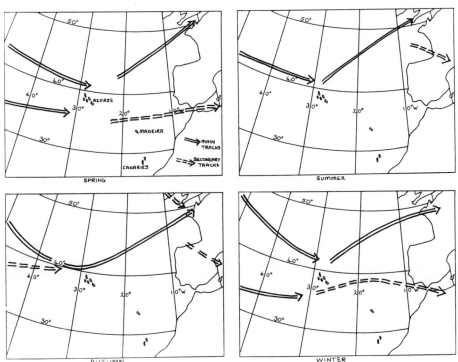

117. Main and secondary tracks of anticyclones

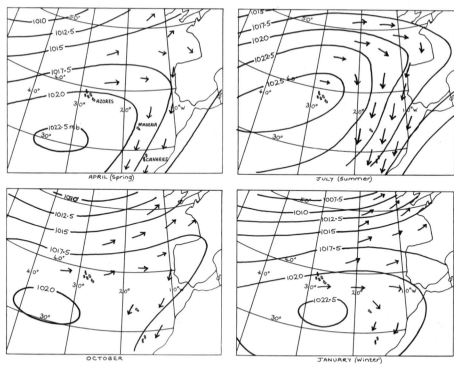

118. **Mean monthly pressure (mb) for the 'seasonal' months**

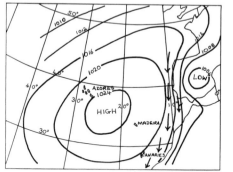

119. **Typical summer surface chart showing the Azores High, a heat low over Spain and the Portuguese Trades**

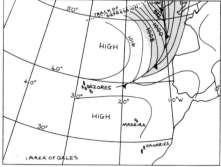

120. **Example of strong pressure gradients on either side of a deeply-troughed cold front**

low over Iberia in summer as it is to the high pressure area to the west, though some of the strength is due to the constraining influence of the often mountainous hinterlands of Iberia and Morocco (Diagram 119).

In general, apart from the Azores route mean speeds gradually decrease southwards (Diagram 121). In

summer, however, the effect of the Portuguese Trades maintains mean winds of around 10 to 15 knots along the whole of the southern routes. In winter winds of 15 to 20 knots in northern waters and along the route to the Azores give way to mean winds of less than 10 knots southward of about 39°N.

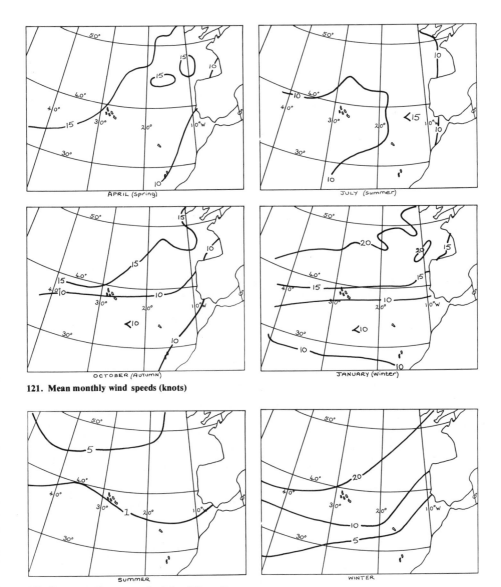

121. Mean monthly wind speeds (knots)

122. Percentage frequency of winds of Force 7 or more, summer and winter

In winter the frequency of winds of Force 7 or more is about 20 per cent in the north and along the route to the Azores, but this decreases to less than 10 per cent along the other routes. In summer a frequency of less than 5 per cent in the north reduces to 1 per cent or even less, towards all the oceanic islands and Gibraltar (Diagram 122).

In spring and summer the frequency of gales (Force 8 or more), is 5 per cent or less; in autumn it is around 10 per cent in the north decreasing to 5 per cent or less in south and southwest. Gales occur on 10 per cent of occasions in winter in northern areas and along the Azores route but this reduces to less than 5 per cent eventually along the

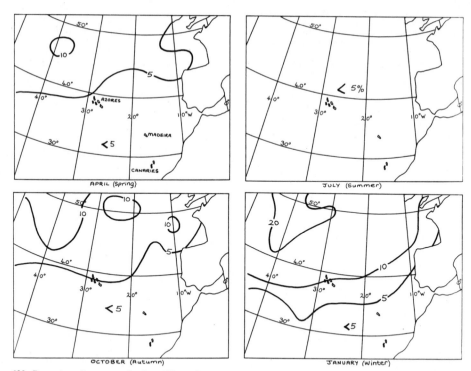

123. Percentage frequency of gales of Force 8 or more

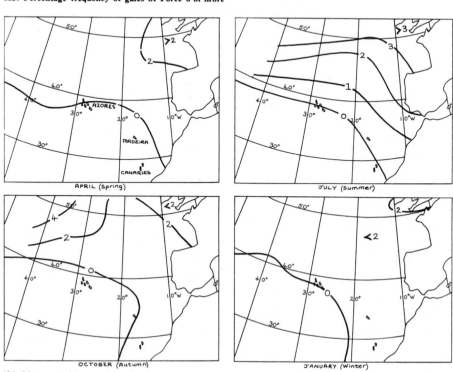

124. Mean monthly percentage frequency of fog

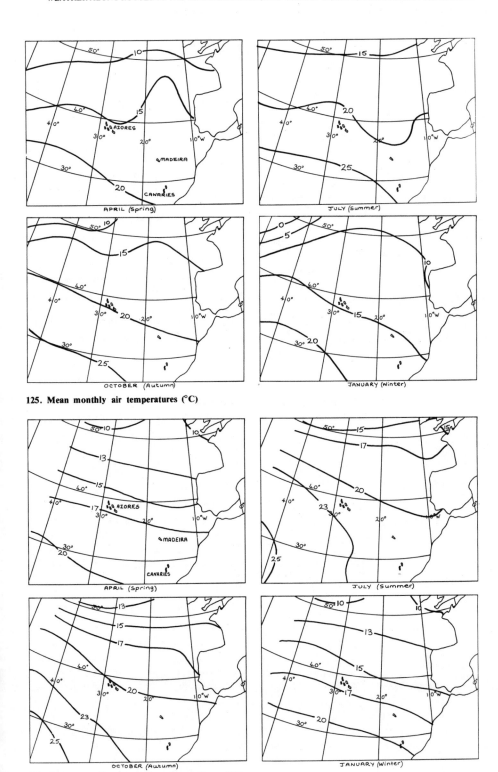

125. Mean monthly air temperatures (°C)

126. Mean monthly sea surface temperatures (°C)

other routes (Diagram 123).

Fog

In each season the risk of fog decreases steadily along the routes with increasing distance from the Southwest Approaches (Diagram 124). Over northern waters the risk is greatest in late spring and summer when sea surface temperatures lag behind air temperatures. Any incursion of warm moist tropical maritime air is liable to lead to fog in this area.

The effect of the cool current setting south along the coast of Iberia and Morocco, called the Canaries Current further south, is clearly seen since at any given latitude the risk of fog increases eastwards towards these coasts; however, the fog risk is still only slight.

In all seasons the percentage frequency is almost zero for the Azores, Madeira and the Canaries but it is higher, especially in summer, in the vicinity of the Straits of Gibraltar.

Air and Sea Temperatures

To amplify the previous discussion on fog and to indicate latitudinal and seasonal warming, mean air and sea temperatures are given in Diagrams 125 and 126. It is noticeable that sea temperatures are lower towards the Iberian and African coasts than they are in mid-ocean. This is due to the south-setting current on the eastern side of the ocean (see Diagram 75) which brings cool water from higher latitudes. Since the air over the oceans is heated by the sea surface, this effect is also seen in the pattern of air temperature distribution. The cooler sea surface also forms a marked inversion of temperature, which often contains a layer of cloud over the southern parts of the route. (The eastern parts of the sub-tropical oceans are typically fairly cloudy, but dry, areas for this reason.)

Cloudiness

In general, cloud amounts decrease with increasing distance from the UK. In the Southwest Approaches skies are clear about 20 per cent of the time and almost completely cloud-covered on about 45 per cent of occasions in most seasons, but in winter the percentages become 15 and 60 respectively. The frequency of cloudy days decreases markedly along the routes to become 25 per cent except towards the Azores where it remains about 40-50 per cent in all season. However, for the reason discussed in the last paragraph, the frequency of completely clear days remains around 20 per cent which is unusually low for sub-tropical latitudes. On the final section of the route to the Mediterranean, clear skies predominate - 70 per cent in summer and autumn and 45 per cent in winter and spring.

Currents

The currents are light and variable over the greater part of the routes, which lie within the great clockwise drift of the North Atlantic Ocean (see Diagram 75). However, the weak southerly set on the eastern side of the ocean gradually strengthens, especially south of about 42°N, and later becomes a major stream, the Canaries Current. Its rate north of about 35°N is probably generally below ½ knot but further south it may attain 1 to 2 knots at times.

Caution

In the vicinity of, and between the individual islands of oceanic groups, the currents may sometimes run at a considerable rate due to the constraining effect of the channels and offlying banks. In extreme cases these rates may reach 10 knots or so.

Oceanic island groups should therefore be approached with caution, especially at night or in poor visibility.

CHAPTER 24

Weather Along the Remainder of the Routes to the West Indies and Caribbean

THE climate along the routes to the West Indies ranges from the warm, settled, dry weather of the sub-tropical high pressure belt in the northeast through the trade wind belt to the hot, humid, thundery weather at the northern edge of the equatorial doldrums in the Caribbean.

Trade wind sailing, typified by blue skies and patchy clouds (trade wind cumulus) and a following Force 4 wind, is perhaps marred only by incessant rolling in the quartering seas. As the westing is run down, cumulus clouds become progressively taller and finally give way to occasional cumulonimbus clouds with their associated violent thunderstorms. Moreover, for the summer half of the year, hurricanes range across the western part of the tropical North Atlantic.

To avoid the hurricane season of late May to early December most yachts leave the Azores, Madeira or the Canaries in early December for the crossing to the West Indies. The passage can be made during the hurricane season (after all, on average, only six will occur each season), but then storm warning broadcast frequencies should be continuously monitored to avoid any possibility of sailing anywhere near a hurricane. Even then, there is no guarantee of avoiding them, as a hurricane has to develop somewhere, perhaps in the vicinity of the yacht, and at a rate which may not allow avoiding action. Since radio failure leaves the yacht in a very vulnerable situation, the safe course is to make the crossing between early December and early May.

Radio Forecasts

The sub-tropical high pressure area with its attendant light winds normally resides across the routes to the West Indies. The position and movement of this anticyclone, together with forecasts of winds and weather along the routes and any other relevant information, is given in the North Atlantic Weather Bulletin and Fleet forecast broadcast from the UK (Diagrams 111 and 112) and also in forecasts from Rota in Spain and Portsmouth, Virginia. Further west, more detailed forecasts may be obtained from numerous stations in the West Indies and USA (Diagram 127). Full details of these broadcasts are given in the Admiralty List of Radio Signals Vol 3, World-wide Marine Weather Broadcasts (USA) and similar publications.

Pressure

The sub-tropical anticyclone dominates over the northern and

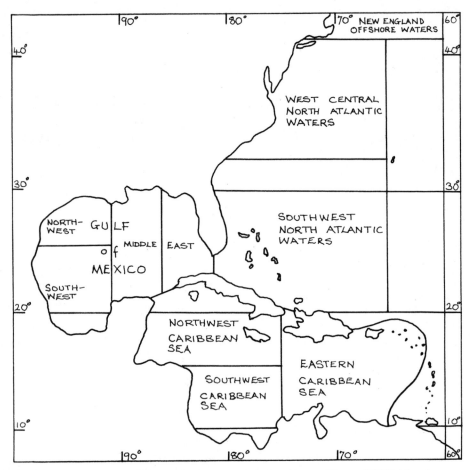

127. Shipping forecast areas in the Caribbean, Gulf of Mexico and **Northern Approaches. Detailed forecasts for American coastal waters,** **including Puerto Rico and the Virgin Islands, are also available – see** **ALRS Vol 3 et al.**

northwestern parts of the routes with its associated light winds or calms and fine dry weather. Pressure decreases southwards and southwestwards. The barometer reading on board should be compared with the values in Diagram 128 after correcting for diurnal variation (Diagram 66) as a constant check on tropical storm development in the vicinity. (The barometer should be set to the sea level value obtained from a nearby meteorological office just before sailing.)

Because of the persistence of the sub-tropical anticyclone to the west of Madeira and the Canary Islands, which

would mean persistent light and variable winds, the normal course is a little southward of the great circle route so that the northeast trades may be picked up as early as possible, and then a course shaped for the West Indies. Since most yachts make the crossing in December and January, the wisdom of shaping a course southward of the great circle is shown in Diagram 128 (d). On setting off from the Azores there is little else to do but follow the shortest track across the anticyclone after listening to weather bulletins to assess where the shortest route will lie. Due to the fact that it normally takes

128. Mean monthly pressure (mb)

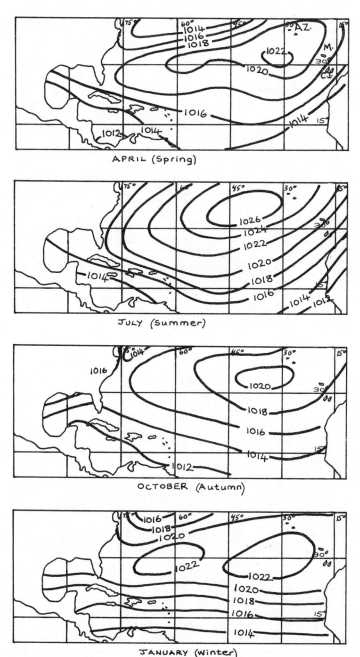

APRIL (Spring)

JULY (Summer)

OCTOBER (Autumn)

JANUARY (Winter)

much longer to pick up the trades on setting off from the Azores than it does from the other oceanic islands, the Canaries and Madeira are more popular departure points.

Winds

The predominant, i.e. most frequent, wind directions along the routes from the Canaries and Madeira are, at first, northeasterly and later east-north-easterly, the true direction of the

189

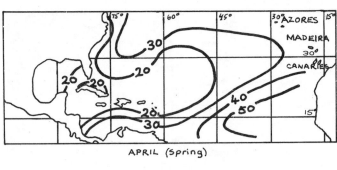

APRIL (Spring)

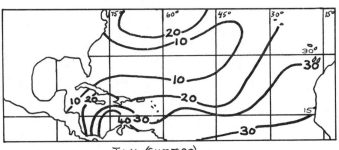

JULY (Summer)

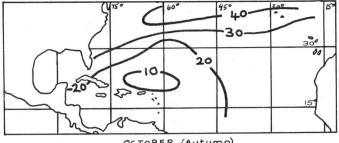

OCTOBER (Autumn)

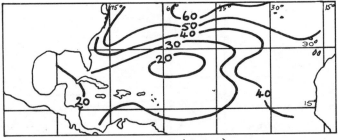

JANUARY (Winter)

129. Percentage frequency of winds of Force 5 or more

northeast trades. The constancy (meaning the frequency of the predominant directions compared with all other directions, normally expressed as a percentage) is at first very low, less 30 per cent due to the wanderings of the Azores anticyclone. It increases to between 60 and 80 per cent over the remainder of the route. These figures apply to the winter season. In the other seasons the constancy of the trades is highest (over 80 per cent along most of

130. Percentage frequency of winds of Force 7 or more

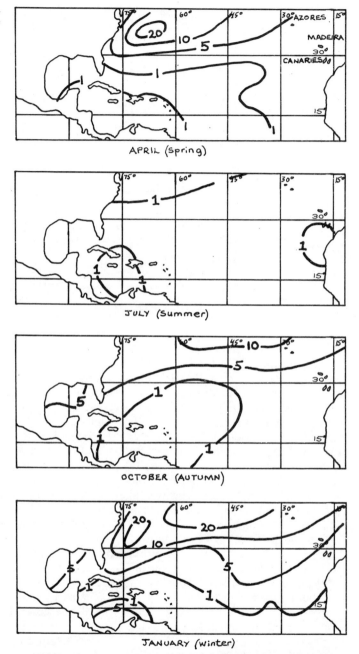

APRIL (Spring)

JULY (Summer)

OCTOBER (AUTUMN)

JANUARY (Winter)

these routes) in summer and least in autumn when it is generally less than 60 per cent. For the first half of the route from the Azores variable winds from almost any direction will normally be experienced in summer and autumn, and strong head winds along the great circle route in winter and spring.

In winter the direction of the trades remains steady at about east-north-east across the ocean and into the Caribbean and Gulf of Mexico. In

191

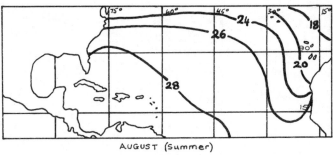

AUGUST (Summer)

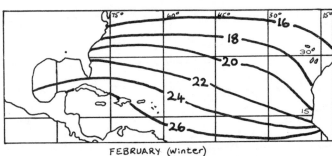

FEBRUARY (Winter)

131. Mean sea surface temperatures (°C)

summer the directions become more easterly beyond about long 60°W and even southeasterly over the Bahamas and northern Gulf of Mexico.

The northern limit of the trades shifts in step with the seasons. It normally lies in about 30°N in summer and about 25°N in winter. The strength of the trades is generally about 15 knots though this too varies with the seasons. They blow more strongly in late winter and spring and are at their weakest in autumn.

The highest frequencies of winds of Force 5 or more, and Force 7 or more, occur in the winter and spring on the route from Madeira and the Canaries, and both indicate the advisability of shaping a course initially just southward of the great circle to the Caribbean. The high frequencies westward of the Azores are associated with southwest to west winds (Diagrams 129,130).

To summarize, the favoured staging posts are Madeira and the Canaries; the best time to begin the crossing is in the winter (December) when the hurricane season is normally over and also to take advantage of the somewhat stronger trades. The initial course should be laid a little southward of the great circle in order to pick up the trades as early as possible.

On the leeward side of many of the islands of the Caribbean the northeast trades normally give way to a southwest or westerly sea breeze during the afternoon, which may typically reach 10 to 15 knots by mid-afternoon.

Thunderstorms are liable to affect the latter halves of the routes and also the Caribbean area in any season. Squalls are often associated with these thunderstorms, their approach being indicated by the characteristic arch or roll described in Chapter 12. Violent squalls may occur when the roll cloud is close overhead.

Another wind effect often initially squally in nature is the Norther of the Gulf of Mexico and, to a lesser extent, of the northern Caribbean. This is normally associated with the passage of a cold front across the area in winter, though due to the low latitudes and to the previous land track over the USA the front will be otherwise weak or sometimes cloudless. However, as it or its remains passes by, the wind typically rises suddenly in a bitterly cold squall from the north and then may persist at gale force, or even stronger, for a day two over the Gulf of Mexico. Due to the sheltering effect from the larger islands and to the warming over the sea, the norther is a much weakened feature by the time it reaches the central Caribbean.

Hurricanes

Tropical revolving storms, their distribution and avoidance are descibed in Chapter 17. It is sufficient, perhaps, to repeat here that yachts and other small craft cannot be expected to survive in the highly chaotic sea conditions close to the centre of a hurricane and ships have been known to founder as well. These storms are therefore to be avoided at all costs. The hurricane season in the North Atlantic lasts from late May to early December, during which period an average of six rage across parts of the western North Atlantic. In bad years there may be as many as 11 hurricanes or in good years only one or two. Most occur during August, September and October. There were five in August 1969 and in

September 1961 again five and four in October 1969. These statistics refer to hurricanes (maximum sustained winds 64 knots or more Force 12). The average number of tropical storms (Force 8 to 11) and hurricanes is ten in a season, but in a bad year this may be 20 or more.

Visibility

Fog at sea is extremely rare along the routes to and within the Caribbean though it may sometimes be met at dawn on the coasts of low-lying areas on the Caribbean islands. However, the visibility may well be reduced to only a few hundred metres in the violent thunderstorms which occur along the latter half of the route and within the Caribbean.

Air and Sea Temperatures

Summer and winter sea temperatures are shown in Diagram 131. Since the air temperatures are controlled by the sea temperature, they will be similar to them at least over the open sea.

Currents

The ocean currents affecting the routes to the West Indies and Caribbean are shown in Diagram 132. On leaving Madeira and the Canaries a favourable current of about 1 knot sets southwest (the Canaries Current) and later turns

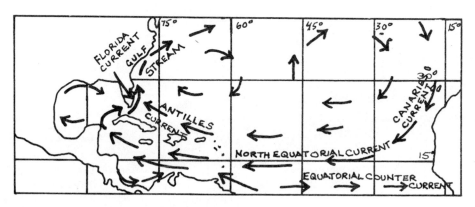

132. Generalized currents of the routes to the Caribbean

westwards in about 15°N as the North Equatorial Current at 1 or 2 knots. A branch of the South Equatorial Current (see Diagram 75) is deflected to run on the north side of the equator by the coast of Brazil. Thus a strong current of 1 to 2 (sometimes 3 to 4) knots sets westward along the north coast of Brazil and floods through the greater part of the Caribbean.

The main exit available for the water accumulating in the Caribbean is the Yucatan Channel between Honduras and Cuba and through this the current flows at a rate of 2 to 5 knots; it is the source region of the Gulf Stream. The current then turns eastward to round the southern tip of Florida and later sets north off that peninsula where it is known as the Florida Current. Here the rates are generally 2 to 4 knots and sometimes 6 knots. Northward of the Bahamas the Florida Current is joined by the Antilles Current the combined flow continuing northeast as the Gulf Stream.

Elsewhere in the general area shown the currents are weak, forming the inner parts of the clockwise circulation(s) of the North Atlantic. The diagram also shows that the currents along the great circle routes from Madeira and the Canaries are weaker than those along the routes just south of the great circle.

Again, the routes from Madeira and the Canaries are therefore more favoured with fair currents (especially south of the great circle) than the route from the Azores which has little or no current along its greater part.

The final approach to the West Indies should be made with caution since the currents in the immediate offing and between the islands often set at a considerable rate, generally in some direction towards the west.

CHAPTER 25

Weather in the Mediterranean

THE climate of the Mediterranean is probably one of the most agreeable to man. It is typified by hot dry summers with mainly light winds and cool wet windy winters. During the winter northerly winds bring cold polar air into the Mediterranean, where it is heated by the relatively warm sea resulting in the formation of vigorous depressions. In the summer season any polar air entering the region is so modified by the European land mass that it is normally warmer than the sea and so depressions are not common in this season. Temperatures apart, it is the ability of northerly winds to generate depressions in the Mediterranean which distinguishes winter from summer.

The two seasons are separated by transition periods which have no special characteristics of their own, unlike spring and autumn in higher latitudes. Rather, the transitions are a mixture of the two main seasons during which time the oncoming season makes several progressively more successful attempts to take over from the old. They normally take place in April and October/November in the east and May and October in the west, where the summers are neither so hot nor so dry as they are in the east. For the purposes of this chapter the transition periods should be understood as April/May and October/November, though in the 'seasonal' diagrams they will normally be represented by April and October.

During the summer the winds are normally light though there are some areas where fresh or even strong winds prevail. In the light wind regions sea and land breezes greatly facilitate coastal passages (it is also worthwhile making a detour to take advantage of these on- and offshore winds). During winter periods of strong winds or gales occur in any part of the Mediterranean.

The greater part of this chapter will be devoted to the winds. In planning a Mediterranean cruise it is well worth studying the local and seasonal winds. Even so there will be periods of frustration due to the fact that the winds normally show considerable variability both in direction and speed.

The air around the region, as elsewhere, is constantly moving under the influence of pressure gradient forces. In winter, when these forces are stronger due to the travelling depressions and anticyclones both within and just outside the area, there is much transfer of air at or near the surface within and across the coasts of the Mediterranean Basin according to the shifting surface pressure patterns. This movement of air is considerably

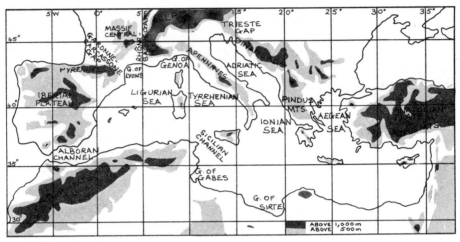

133. Topography of the Mediterranean region

obstructed and modified by the topography of the region.

Diagram 133 shows that almost the whole of the Mediterranean basin is surrounded by high ground, above 500 m. Much of this extends above 1000 m and in some ranges, such as the Atlas, Taurus and Pindus Mountains and the Apennines, Alps and Pyrenees, occasional peaks rise above 3000 m. The only major part of the Mediterranean coastline which is not backed by mountains is that of Egypt and the greater part of Libya.

In winter, depressions normally cross the region from west to east, each generally followed in turn by an anticyclone on the remote side of the northern mountains. This, naturally, provides a pressure gradient across the mountains. The relatively narrow gaps and passes within and between the mountain ranges now assume tremendous importance for the wind is funnelled through them sometimes at a considerable speed, and often subsequently retains its new direction and much of its speed for several hundreds of miles across the sea. These winds have been infamously known since the earliest civilizations in the Mediterranean: not surprisingly, they have long since acquired names many

of which are well-known even away from the Mediterranean area. Most of the named winds are entirely local effects; others, however, have no particular characteristics to distinguish them from winds of similar direction in other parts of the world, but have acquired names simply because of ancient custom. The former will receive fairly close attention here; the latter only brief mention.

Local Winds

The main local winds of the Mediterranean are shown in Diagram 134. Perhaps the best known of these is the Mistral.

The Mistral occurs when pressure is higher to the north and west of the Massif Central, Alps etc than it is over the western Mediterranean. The most typical development of the Mistral follows the eastward movement of a low on the northern side of the mountains, whose cold front eventually breaks through into the Mediterranean Sea. In winter this normally leads to the generation of a vigorous depression (or more correctly, to the rapid deepening of a weak 'lee' low) in the Gulf of Genoa. Meantime, pressure has been rising over northern France so that a considerable pressure

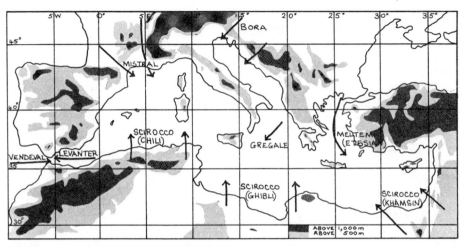

134. The main regional winds of the Mediterranean

gradient occurs across the mountain barriers. Most of this is channelled into the Rhone-Saone and Garonne-Carcassonne Gaps (see Diagram 134) through which the air then flows at a tremendous rate. These 'gap' winds in this area are known as the Mistral. An example of the main synoptic situation producing the Mistral is given in Diagram 135.

The onset of the Mistral is often dramatically sudden. Arriving as it does in association with a cold front, it often introduces itself as a violent squall, but unlike a normal squall there is no early let-up. Further squalls may

occur from time to time, especially in association with further cold troughs arriving in the area. If the Genoa low is slow-moving or moves southeast, the Mistral may persist as a strong to gale northwesterly beyond Corsica and Sardinia to the Sicilian Channel (where there is a local increase due to funnelling) and even sometimes as far as Malta. It sometimes arrives at the Balearics as a north-northeasterly wind. The Mistral is normally most powerful in the winter season when the sea is still relatively warm and its northerly winds may bring very cold air into the region. In summer the Mistral

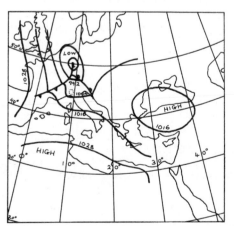
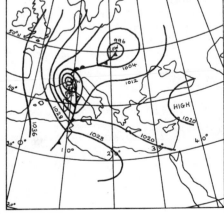

135. The development of a favourable pattern for the Mistral

Speed kts	Jan	Feb	Mar	Apr	May	June	July	Aug	Sep	Oct	Nov	Dec	Year
>21	10	9	13	11	8	9	9	7	5	5	7	10	103
>27	8	6	10	9	6	4	4	3	2	2	3	7	64
>33	4	4	6	5	3	2	0.6	1	0.6	0	0	4	30
>40	2	2	3	0.5	0.7	0.5	0.3	0.5	0.3	0	0	1	11

Table 9. Mean number of days with strong Mistral

is less powerful and less common. It can attain Force 8 and on rare occasions will attain Force 10 close to the French coast, especially just downwind of the main gaps.

The strength of the Mistral is sometimes reduced during the afternoon by the sea breeze effect in late winter and in summer, when daytime air temperatures rise sufficiently high above the sea surface temperatures to produce a sea breeze (see Chapter 12). This can only occur in moderate or weaker Mistral (less than about Force 7). However, the katabatic effect (Chapter 12) probably invariably accelerates the flow of air through the gaps, and so it is likely that the Mistral will be stronger at night.

Both the sea breeze and katabatic effects are greatest with clear skies which in any case normally accompany the arrival of the Mistral, at least along French coasts; further east the deepening of the Genoa low produces much thick cloud and rain. The sharp clearance of cloud and rain from the north or northwest at the French coast often heralds the onset of the Mistral over the sea. The barometer, however, will probably continue to fall due to the deepening of the Genoa low.

The Bora is another gap wind. It occurs in the Adriatic region when pressure is higher over and to the north of the Hungarian Plain than it is over the Adriatic and western Mediterranean (Diagram 136). Strong winds then occur through the gaps and passes in the Dinaric Alps and arrive over the Adriatic as strong northeasterly winds. The best example of Bora occurs through the Trieste Gap where it sometimes drives across the Adriatic to Venice. When the Bora is widespread over the Dinaric Alps it commonly becomes a northwesterly wind over the southern Adriatic though on some occasions it maintains its northeasterly direction across the whole of this sea area and even over the Apennines of Italy.

Though the Bora may occur at any time of year it is more frequent in winter when the required pressure pattern is more likely to occur.

As with the Mistral, its onset is often very sudden. It also often begins as a violent squall associated with the passage of a cold front into the head of the Adriatic.

The Bora often reaches Force 7 to 8, and sometimes much more. It is then dangerous to yachts who may find themselves being driven towards the lee shore of northeast Italy where there is

Jan	Feb	Mar	Apr	May	June	July	Aug	Sep	Oct	Nov	Dec	Year
8	6	4	2	1	0.4	0.8	1	2	3	5	6	39

Table 10. Average number of days of Bora at Trieste

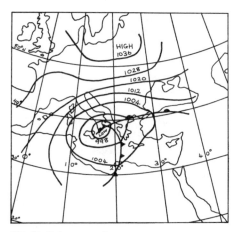

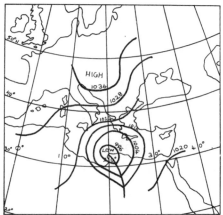

136. Typical synoptic situation producing Bora winds in the Adriatic

137. Typical situation producing Gregale winds in the Ionian Sea

little shelter. At Trieste the highest gust recorded, in Bora conditions, was 110 knots (the associated mean hourly wind speed was 70 knots). Over the open sea the speed is less than over land at the exist of the gaps. The average duration of a Bora gale at Trieste is about 12 hours but it can last for two days; the duration of a Bora which only sometimes reaches gale force is about 40 hours.

The Gregale is a strong northeasterly which results from high pressure over the Balkans and low pressure over the southern Mediterranean (Diagram 137). Air is then forced to flow through the gaps and passes of the Dinaric Alps and Pindus Mountains and then streams southwest across the Ionian Sea and Malta. It also occurs with the passage of a low eastwards along the southern Mediterranean when the mountains of southern Italy and Sicily force the air on the northern flank of the low to turn towards the northeast.

The Gregale commonly reaches Force 7 to 8 and may sometimes reach storm Force 10. It is clearly more frequent in winter than in summer.

The Meltemi or Etesian is the only local wind of the Mediterranean region which prevails throughout the greater

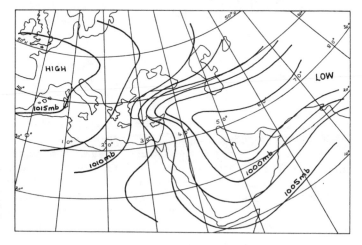

138. The 'monsoon' trough of the east Mediterranean from the Southeast Asia heat low

	Jan	Feb	Mar	Apr	May	June	July	Aug	Sep	Oct	Nov	Dec	Year
Maximum number of spells	4	3	2	2	1	1	0	1	1	2	3	3	23
Average number of spells	1.6	1.1	0.7	0.7	0.2	0.2	0	0.2	0.2	0.5	0.8	0.9	7.1
Average number of days with wind 35 to 43 knots	0.4	0.9	0.3	0.1	0	0	0	0	0.1	0	0	0.5	2.3
Average number of days with wind 43 to 50 knots	0.1	0.3	0	0	0	0	0	0	0	0	0	0	0.4

Table 11. Gregale frequency at Malta. (In all months the minimum of Gregale spells is 0.)

soons of southeast Asia. As explained in Chapter 2, the Siberian anticyclone of winter gives way in summer to a large area of low pressure (a heat low) over southeast Asia; the flow around its southern and eastern flanks being the Southwest Monsoon of the China and Arabian Seas and the Bay of Bengal. A trough of low pressure extends from the southeast Asia heat low across Afghanistan and Iraq towards the northeast Mediterranean (Diagram 138). At the same time there is normally higher pressure over and just to the north of the Black Sea resulting in northeasterly airflow across Turkey. This itself produces a lee trough along the south coast of Turkey which effectively extends the monsoon trough westwards as far as the Aegean (Diagram 139), and results, because of the Anatolian Plateau, in a strong flow of air from the Black Sea area through the Dardanelles gap and south through the Aegean, later emerging southeastwards as a weaker flow.

The Meltemi normally begins in June and ceases in September. Its speed usually increases to Force 6 or 7 (occasionally Force 8) during the afternoon and evening and normally decreases a little during the night. Its direction is northeast in the north, north in the centre and more northwest in the southeastern Aegean; at Crete it is northerly. Wind directions in summer in the channels between Crete and Kithera are usually westerly.

When planning an Aegean cruise it is normal to make one's northing early in the season, April or May, when the winds may often be more favourable and then to cruise through the islands making a slow southing when the Meltemi is established. When sailing towards the Aegean from the west, a light westerly wind may be expected, while from the east, the Levant coast and Cyprus, a light to moderate west-northwesterly is to be expected in summer; the main alternative is calm.

part of a season. (The Turkish word 'Meltemi' is now widely adopted in the Aegean area in place of the Greek 'Etesian'.) The Meltemi is therefore 'monsoonal' in character and this is not too inappropriate a description as it results from the same global atmospheric disturbance as the mon-

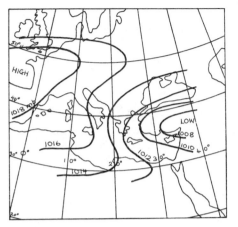

139. Typical summer pressure pattern produc- ing the Meltemi of the Aegean

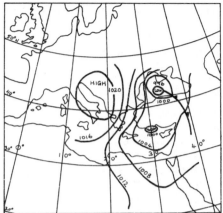

140. Typical synoptic situation associated with strong northerly winds in the Aegean in winter

Within the Aegean the Meltemi prevails with great constancy from June to September. However, weak depressions sometimes cross the Aegean area in summer, temporarily destroying it over a part, if not the whole of the Aegean. A fall of pressure below the seasonal norm (see Diagram 151b) may indicate such an interruption is about to take place. The recovery of pressure after the passage of the weak low will only briefly presage the re-establishment of the Meltemi.

A quite different and often much stronger 'northerly' occurs in the Aegean in winter when a depression passes across the southern Aegean and then over or around Turkey. As the low passes by high pressure normally becomes quickly established over the Balkans and Adriatic (Diagram 140). The scene is now set for the channelling down the Aegean which readily occurs, resulting on many occasions in winter in violent winds of Force 9 or more. With particularly strong northeasterly gradients over the Aegean the wind may blow at Force 10 or more in the Kithera channels and even more strongly around the headlands on the southern side of the Peloponnesus.

Unlike the pure Meltemi, these northerly winds of winter normally last a day or so and sometimes for three to five days.

The Scirrocco is a hot, initially dry, wind from the deserts of North Africa and Saudi Arabia. The name Scirrocco, of Italian origin, is more commonly adopted in the northern countries but in the south it is known by different names. From Morocco to Tunisia it is called **Chili**, in Libya the **Ghibli**, and in Egypt and now in Israel **Khamsin**. The Scirrocco normally occurs in advance of a depression moving eastward across the Mediterranean region. Its chief characteristics are the dust it carries and its eventual high humidity, both extremely unpleasant. It is dry or moist depending on its recent sea track and speed.

The Scirrocco is always very dry on leaving the coast but because of its high temperature it quickly picks up moisture in its lower levels (below about 1 km) once over the sea. The amount of moisture acquired depends on the time the air has spent over the sea; a relatively low wind speed over a long sea track (say, Force 3 from Gulf of Sirte to Sicily) will mean a very moist Scirrocco, probably with extensive stratus cloud, drizzle and associated poor visibility. A stronger wind speed over a short sea track (say Force 6 from Tunisia to Sicily) will mean a dry Scirrocco. The latter, however, will carry much more dust which may reduce visibility to a few kilometres and sometimes to a few metres.

Since depressions are more frequent in the Mediterranean in winter, the Scirrocco is a winter wind. In the eastern Mediterranean depression frequency reaches a peak in late winter and in the spring transition. Khamsin average frequency at Alexandria is four to five days each month from March to May, one to two in January, February and September, and less than one in the other months. The Scirrocco (Ghibli) is more frequent over Libya than is the Khamsin over Egypt.

Diagram 141 shows a typical example of a Scirrocco over Cyrenaica and the Ionian Sea. The low, of desert origin, has moved northeast from Tunisia. The strong southerlies ahead of the low raise considerable dust and sand seriously reducing visibility over sea areas where these winds prevail. The passage of the cold front clears the dust.

The Levanter is an easterly wind through the Alboran Channel and Straits of Gibraltar. It occurs at any time of year when pressure is relatively high over the Iberian Peninsula or when there is a low to the south or

141. Typical synoptic situation producing strong Scirrocco (Ghibli) over Cyrenaica, the Ionian Sea and the southern Adriatic

southwest of Iberia. In summer the Levanter is moist and produces fog patches in the Straits and its eastern approaches. In winter the Levanter is more often associated with a depression passing close southward of Gibraltar and is then accompanied by thick cloud and often heavy rain. It may blow in excess of Force 8 for days on end between October and March. In the summer months the maximum wind is typically Force 7.

The Vendeval is a westerly wind in the same area, resulting from a low passing over Iberia. It is more frequent in winter (when it may exceed Force 8) than in summer.

142. Annual percentage frequency of gales of Force 8 or more over the open sea

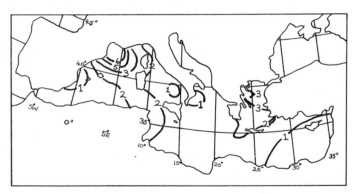

143. Annual percentage frequency of winds of Force 6 or more over the open sea

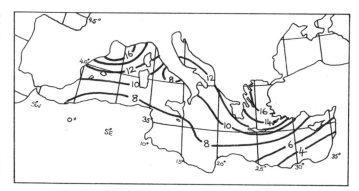

144. Most frequent directions of winds of Force 6 or more

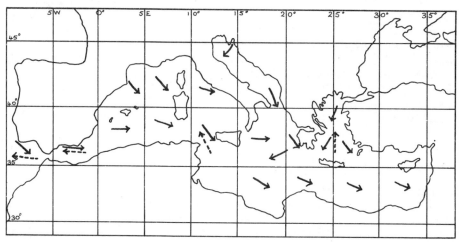

The Leveche is the name given to a Scirrocco on the south and southeast coasts of Spain. Due to the short sea track it is normally dry but dusty. The Leveche may attain gale force at times.

The Marin is the name given to the Scirrocco in the Gulf of Lyons. It normally arrives when a depression is forming in the west Mediterranean and therefore is typically accompanied by thick cloud and rain.

The Libeccio is a west to southwest wind over Italian sea areas; it is commonly associated with depressions in the Gulf of Lyons.

The Levante is a wind of long fetch from about northeast blowing across the western Mediterranean on to the east coast of Spain. It may sometimes exceed gale force.

The Tramontana is a mainly northerly wind blowing across Italy to the Tyrrhenian Sea.

The Grecale is a northeasterly wind blowing across Italy and Tyrrhenian Sea. (The name is Italian, and not to be confused with the Maltese Gregale, described above.)

The Maestro is a name commonly given to the northwesterly in the Adriatic in summer.

Mediterranean Wind in General

Not surprisingly, the highest frequencies of winds of Force 8 or more are found in the northwest Mediterranean (over 6 per cent) and in the north Aegean (over 3 per cent), shown in Diagram 142. The pattern is reflected in the annual percentage frequency of winds of Force 6 or more (Diagram 143). The main wind directions associated with these Force 6 winds are shown in Diagram 144.

In the Straits of Gibraltar and the Alboran Channel most strong winds are westerly (Vendeval), though from April to June the easterlies (Levanter) are more frequent.

In the western Mediterranean, including the Tyrrhenian Sea and the Sicilian Channel, the main strong wind direction is northwesterly (Mistral) though southerly or southeasterly (Scirrocco) may, less frequently, produce gales in the region.

Most strong winds in the north Adriatic are northeasterly (Bora) and northwesterly to westerly in the southern half of this sea and over the Ionian Sea, though in the latter region the Gregale (northeasterly) is of about equal frequency.

In the Aegean the main strong wind direction is northeasterly in the north, through northerly to northwesterly in the south. In winter the winds are stronger than in the summer (Meltemi). Strong southerly winds occur in the Aegean in winter ahead of lows about to cross the region. In the Aegean area, in summer, a fall of pressure (beyond diurnal variation) will normally mean a weakening of the Meltemi. In winter it may precede a southerly gale as a depression approaches from the west. In both seasons a rise of pressure (after a fall) will be followed, often very quickly, by northerly winds. In summer this means the reinstatement of the Meltemi while in winter it may lead to violent northerly winds of storm force.

Over the eastern Mediterranean, east of a line from Rhodes to Crete to Cyrenaica, the main strong wind directions are west to northwest. High pressure over Turkey in winter may produce strong or even gale force easterly winds over the northeast of this area including the Cyprus region. Rising pressure in winter, reaching several millibars beyond the seasonal

normal, is usually a good indication that easterly winds will follow; the greater the rise the stronger the easterly. Subsequent falling pressure normally precedes a return to northwest to west winds.

High pressure over Syria sometimes leads to a strong Khamsin (east to southeast) over Israel, though this is an infrequent event mainly confined to late winter and the April/May transition.

Despite the foregoing description of strong winds and gales, light to moderate winds prevail in the Mediterranean for most of the time. This is true even in winter when strong winds reach their highest frequency (Diagram 145).

In winter (January) the percentage frequencies of strong winds varies from over 30 per cent in the Gulf of Lyons to less than 10 per cent in the extreme southeast of the Mediterranean; i.e. on more than 90 per cent of occasions in the southeast and over 60 per cent in the northwest the winds are Force 5 or less!

In the spring transition (April) strong winds are less frequent so that light to moderate winds may be expected over 90 per cent of the time in eastern and southern areas and on more than 70 per cent even in the northwest.

In July, apart from the Aegean, light to moderate winds prevail for over 90 per cent of the time and even in the Aegean the percentage frequency of winds of Force 5 or less is still over 80 per cent.

During the autumn transition (October) strong winds become slightly more frequent in the northwest and definitely so in the Aegean. Even so, over the whole Mediterranean area light to moderate winds prevail for over 70 per cent of the time and over the greater part of the Mediterranean for more than 80 per cent of occasions.

145. Monthly percentage frequency of winds of Force 6 or more

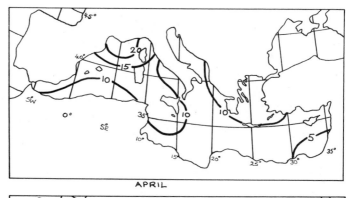

APRIL

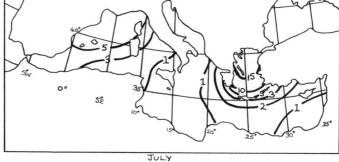

JULY

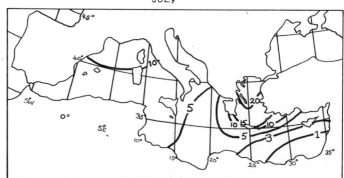

OCTOBER

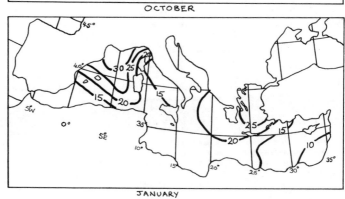

JANUARY

Radio Forecasts

Forecast information for the many parts of the Mediterranean region is available from several nations and Diagrams 146 to 150 show the areas covered. Details of transmission times and frequencies are given in Admiralty List of Radio Signals Vol 3 and World-wide Marine Weather Broadcasts (USA), and from 1983 in *Reed's Nautical Almanac* Mediterranean edition. In addition, forecasts for the waters around Cyprus are available on 2700 kHz at 1133 and 2323 GMT.

146. Shipping forecast areas – UK Fleet forecasts					
1 Nelson	5 Lions	9 Bonny	13 Centaur	17 Sidra	21 Matruh
2 Alboran	6 Unicorn	10 Venice	14 Boot	18 Aegean	22 Delta
3 Valencia	7 Bougie	11 Volcano	15 Malta	19 Jason	23 Taurus
4 Oran	8 Genoa	12 Gabes	16 Ionian	20 Bomba	24 Crusade

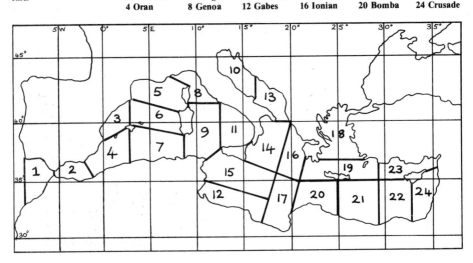

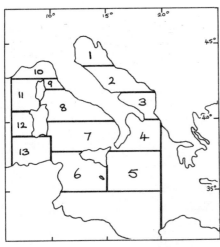

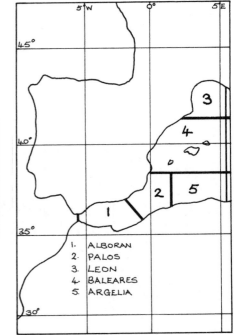

147. Shipping forecast areas – Italy	
1 Alto Adriatico	7 Basso Tirreno
2 Medio Adriatico	8 Medio Tirreno
3 Basso Adriatico	9 Alto Tirreno
4 Alto Ionio	10 Mar Ligure
5 Basso Ionia	11 Mar di Corsica
6 Canale di Sicilia	12 Mar di Sardegna
	13 Canale di Sardegna

1. ALBORAN
2. PALOS
3. LEON
4. BALEARES
5. ARGELIA

148. Shipping forecast areas – Greece

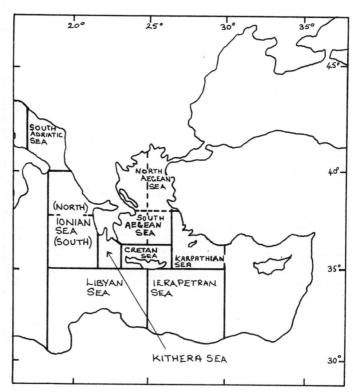

149. Shipping forecast areas – France

1 Alboran
2 Sud Baléares
3 Nord Baléares
4 Lion
5 Provence
6 Ouest Sardaigne
7 Sud Sardaigne
8 Gênoa
9 Est Corse
10 Est Sardaigne

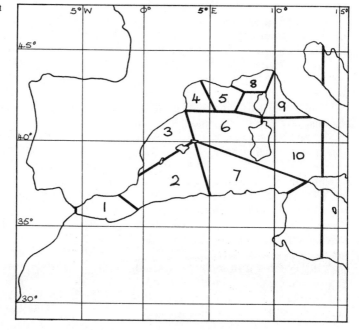

Left:
150. Shipping forecast areas – Spain

Pressure

The distribution of surface pressure over the Mediterranean is given by seasons in Diagram 151. In winter (January) the Mediterranean is normally an area of relatively low pressure. (The shape of the 1015 mb isobar reflects a main depression track from the Gulf of Genoa towards the Aegean, rather than a depression normally centred in the Ionian Sea.) In summer (July) pressure is relatively high in the western Mediterranean and low in the east (the 'monsoon trough'). The charts for the transition periods show that there is normally a slack pressure gradient over the Mediterranean, suggesting light winds, which indeed is usually the case.

Depressions

Few depressions affect the Mediterranean in summer and those which do are chiefly to be found moving east across northern areas producing a moderate Mistral or Bora or reinforcing the Meltemi according to its position. A shallow depression may sometimes be located between Cyprus and southern Turkey in the monsoon trough. Most depressions affecting the Mediterranean occur in winter.

There are two main formation regions, the Gulf of Genoa and the region to the southeast of the Atlas Mountains (Diagram 152). Both regions lie in the lee of predominantly northwesterly winds and so, at least initially, the depression are lee lows. The lows in both regions may later deepen into vigorous circulations. Depressions entering the Mediterranean, invariably travelling west to east, do so either through the Garonne-Carcassonne gap or through the Straits of Gibraltar, though an occasional low may survive the struggle of crossing the Iberian Plateau, its remains being reinvigorated off the east coast of Spain by the lee effect.

Though the tracks of individual depressions generally are determined by the airflow at levels well above the surface (in fact, above 5,000m), the topography of the land surface, especially in the mountainous Mediterranean region, is also a 'steering' factor. Temperature, too, is important in that winter lows tend to remain over the warmer sea surface as long as possible. The tracks shown in Diagram 152 reflect these points reasonably well in that the lows generally move parallel to the mountain ranges, crossing them only when the upper air steering factor assumes overriding influence; most typical tracks remain over the sea.

Depressions also occur in the transition periods. In April/May those affecting the northern areas tend to be less vigorous than in winter, whereas the desert lows in the south are at their most vigorous and generally produce the strongest Scirroccos, with most dust, in the early transition period. In October the reverse is the case. Desert lows tend to be weaker and less frequent whereas north Mediterranean lows may become vigorous and produce gale or sometimes storm force winds in northern sea areas.

In general, the most vigorous depressions occur in winter when the air arriving in the region from some northerly point will be much colder than the sea surface. This contrast of temperature is an essential precondition for the development of new lows or for the deepening of existing lows. The mean annual range of pressure across the whole region is only about 15 mb (Diagram 151). Since Mediterranean lows, though sometimes vigorous, are not as deep as Atlantic lows, nor are anticyclones over the Mediterranean Sea nearly as intense as Siberian anticyclones, the extreme range of pressure over the sea area is from about 970 to 1030 mb.

151. Mean monthly surface pressure (mb)

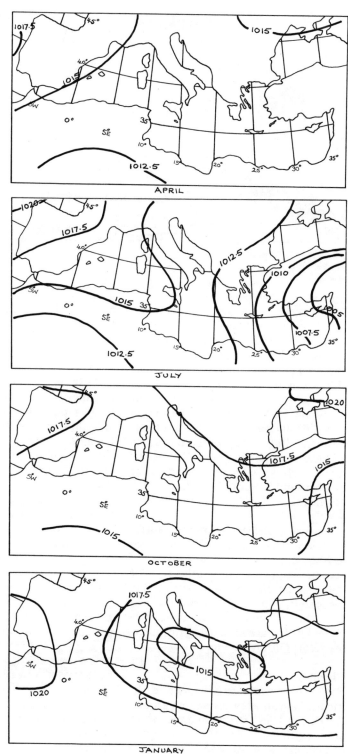

APRIL

JULY

OCTOBER

JANUARY

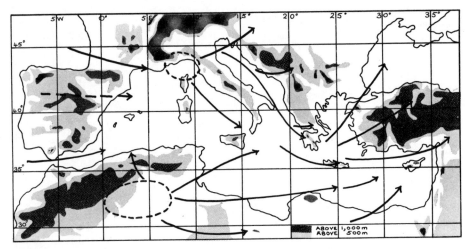

152. Depression tracks and the main areas of formation

Air and Sea Temperatures

The mean annual range of temperature, from summer to winter, is about 12°C. Winter (January) average temperatures range from about 9°C/48°F in northern areas (7°C in the far north of the Aegean) to 16°C/61°F in southeastern areas (Diagram 153). In some winters snow may fall at sea level as far south as southern Italy and southern Greece and exceptionally in Crete. During the summer (July) air temperatures range from around 22°C/72°F in the extreme northwest to 27°C/81°F in the extreme east.

The sea surface temperature gradient in all seasons (Diagram 154) runs from lowest values in the northwest to highest values in the southeast. The lower temperatures in summer (July) in the eastern approaches to the Straits of Gibraltar account for the higher frequency of fog accompanying Levanters (easterlies) in that area.

Visibility

Fog is not nearly so frequent in the Mediterranean as in the waters around the British Isles. The main areas which are sometimes affected are the Straits of Gibraltar and Alboran Channel, the northern sea areas and the eastern part of the Mediterranean. The average number of days of fog at Gibraltar are:

May 2 June 5 July 11 Aug 8
Sept 4 Oct 3 Nov 2

Sea fog is most likely to occur when the easterly levanter prevails in the summer months. In these months the sea surface temperature is relatively low in the Gibraltar area so that the warm moist Levanter winds are readily cooled to their dew-point to form sea fog which may sometimes last for two or three days.

In the northern Mediterranean, such as the Gulfs of Lyons and Genoa and the head of the Adriatic and Aegean, fog is likely to occur in winter and spring when Scirrocco conditions prevail, i.e. warm, subsequently moist southerly winds, usually ahead of an eastward moving depression. The cool sea temperatures in these areas in winter and spring provide the required conditions for sea fog with Scirrocco winds.

Sea fog sometimes occurs in the eastern Mediterranean in spring or early summer when sea temperatures may still be relatively cool though the air is warm, and since it normally moves slowly around this region in spring and summer it is also moist. Sea fog, when

153. Mean monthly air temperatures (°C)

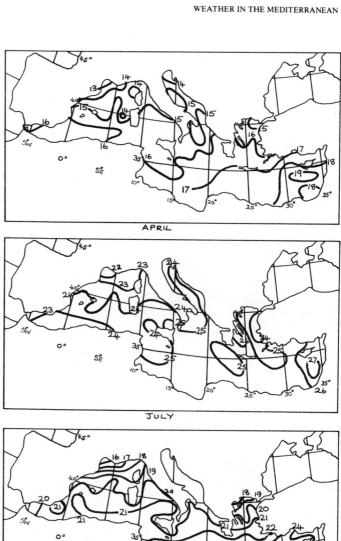

APRIL

JULY

OCTOBER

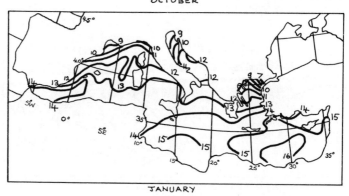

JANUARY

APRIL

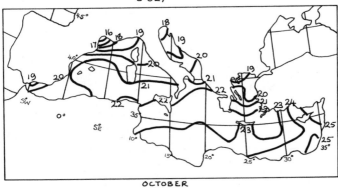

JULY

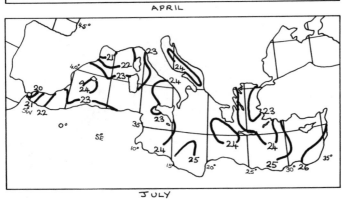

OCTOBER

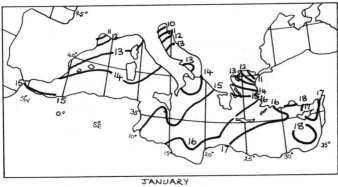

JANUARY

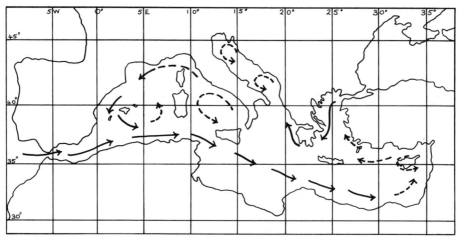

155. Main surface current circulation

it does occur, is patchy.

The visibility may be reduced to around 2 nautical miles in thunderstorms, and sometimes less than 1 mile in violent storms - more likely to occur in winter. Again, in winter, visibillity may be reduced to 1 to 2 miles or less in frontal rain, normally ahead of and near depression centres.

One more factor accounting for poor visibility in the Mediterranean is dust, chiefly from North Africa and therefore of significance under Scirrocco conditions. With strong to gale force south to southeast winds, visibilities near some African coasts are seriously reduced to less than 1 mile for several miles at sea, and perhaps up to 100 miles out. These are rare events and even so are more likely to occur in spring when the strongest and most frequent Scirroccos occur.

Despite these comments, visibility in all parts of the Mediterranean is mainly good and sometimes excellent.

Currents

The net flow of surface water through the Straits of Gibraltar is from the west. This is compensated by a return bottom flow of more saline and dense Mediterranean water. The east-going surface inflow gives rise to the only significant current of the region which is an easterly set through the Straits of Gibraltar and along the coasts of Algeria and Tunisia (Diagram 155). This easterly set while averaging ½ to ¾ knot, may sometimes attain 2, and very rarely 3 knots or more in the Straits and the Alboran Channel. Some vestige of this flow remains across the central Mediterranean (at ¼ to ½ knot) but farther east it becomes very weak.

In general, as Diagram 155 indicates, the currents tend to flow in an anticlockwise fashion around the western, central and eastern Mediterranean and in the Adriatic and Aegean and in most of these areas they are extremely weak.

SECTION 7

Appendices

Appendix 1

PRESSURE

Since pressure is force per unit area, the atmospheric pressure at a place is the weight of the column of air throughout the whole depth of the atmosphere over a given area at that place. The weight of the column is almost always changing and so the atmospheric pressure is also very variable.

The average (global) atmospheric pressure at the earth's surface is commonly quoted as about 14lb/sq in (approx 1 kg/sq cm) or, more appropriately here, 1013 mb; the units used internationally in meteorology. The millibar (mb) is derived from an earlier metric system which used the dyne as the unit of force whereby the average atmospheric pressure is about 1,000,000 dynes/sq cm. Since the dyne/sq cm was too small for meteorological use, 1,000,000 dynes/sq cm was called 1 bar, and of course 1 millibar is one-thousandth of a bar. (In the S.I. system of units, the unit of pressure is a pascal which is 1 newton/sq m. 1 millibar equals 100 pascals.)

Appendix 2

TABLE FOR CONVERTING BAROMETRIC READINGS IN INCHES INTO MILLIBARS

(Equivalents in Millibars of Inches of Mercury at 0°C and Standard Gravity 980·665 cm./sec.²)

Mercury Inches	·00	·01	·02	·03	·04	·05	·06	·07	·08	·09
					Millibars					
27·0	914·3	914·7	915·0	915·3	915·7	916·0	916·4	916·7	917·0	917·4
27·1	917·7	918·1	918·4	918·7	919·1	919·4	919·7	920·1	920·4	920·8
27·2	921·1	921·4	921·8	922·1	922·5	922·8	923·1	923·5	923·8	924·1
27·3	924·5	924·8	925·2	925·5	925·8	926·2	926·5	926·9	927·2	927·5
27·4	927·9	928·2	928·5	928·9	929·2	929·6	929·9	930·2	930·6	930·9
27·5	931·3	931·6	931·9	932·3	932·6	933·0	933·3	933·6	934·0	934·3
27·6	934·6	935·0	935·3	935·7	936·0	936·3	936·7	937·0	937·4	937·7
27·7	938·0	933·4	938·7	939·0	939·4	939·7	940·1	940·4	940·7	941·1
27·8	941·4	941·8	942·1	942·4	942·8	943·1	943·4	943·8	944·1	944·5
27·9	944·8	945·1	945·5	945·8	946·2	946·5	946·8	947·2	947·5	947·9
28·0	948·2	948·5	948·9	949·2	949·5	949·9	950·2	950·6	950·9	951·2
28·1	951·6	951·9	952·3	952·6	952·9	953·3	953·6	953·9	954·3	954·6
28·2	955·0	955·3	955·6	956·0	956·3	956·7	957·0	957·3	957·7	958·0
28·3	958·3	958·7	959·0	959·4	959·7	960·0	960·4	960·7	961·1	961·4
28·4	961·7	962·1	962·4	962·8	963·1	963·4	963·8	964·1	964·4	964·8
28·5	965·1	965·5	965·8	966·1	966·5	966·8	967·2	967·5	967·8	968·2
28·6	968·5	968·8	969·2	969·5	969·9	970·2	970·5	970·9	971·2	971·6
28·7	971·9	972·2	972·6	972·9	973·2	973·6	973·9	974·3	974·6	974·9
28·8	975·3	975·6	976·0	976·3	976·6	977·0	977·3	977·7	978·0	978·3
28·9	978·7	979·0	979·3	979·7	980·0	980·4	980·7	981·0	981·4	981·7
29·0	982·1	982·4	982·7	983·1	983·4	983·7	984·1	984·4	984·8	985·1
29·1	985·4	985·8	986·1	986·5	986·8	987·1	987·5	987·8	988·1	988·5
29·2	988·8	989·2	989·5	989·8	990·2	990·5	990·9	991·2	991·5	991·9
29·3	992·2	992·6	992·9	993·2	993·6	993·9	994·2	994·6	994·9	995·3
29·4	995·6	995·9	996·3	996·6	997·0	997·3	997·6	998·0	998·3	998·6
29·5	999·0	999·3	999·7	1000·0	1000·3	1000·7	1001·0	1001·4	1001·7	1002·0
29·6	1002·4	1002·7	1003·0	1003·4	1003·7	1004·1	1004·4	1004·7	1005·1	1005·4
29·7	1005·8	1006·1	1006·4	1006·8	1007·1	1007·5	1007·8	1008·1	1008·5	1008·8
29·8	1009·1	1009·5	1009·8	1010·2	1010·5	1010·8	1011·2	1011·5	1011·9	1012·2
29·9	1012·5	1012·9	1013·2	1013·5	1013·9	1014·2	1014·6	1014·9	1015·2	1015·6
30·0	1015·9	1016·3	1016·6	1016·9	1017·3	1017·6	1017·9	1018·3	1018·6	1019·0
30·1	1019·3	1019·6	1020·0	1020·3	1020·7	1021·0	1021·3	1021·7	1022·0	1022·4
30·2	1022·7	1023·0	1023·4	1023·7	1024·0	1024·4	1024·7	1025·1	1025·4	1025·7
30·3	1026·1	1026·4	1026·8	1027·1	1027·4	1027·8	1028·1	1028·4	1028·8	1029·1
30·4	1029·5	1029·8	1030·1	1030·5	1030·8	1031·2	1031·5	1031·8	1032·2	1032·5
30·5	1032·8	1033·2	1033·5	1033·9	1034·2	1034·5	1034·9	1035·2	1035·6	1035·9
30·6	1036·2	1036·6	1036·9	1037·3	1037·6	1037·9	1038·3	1038·6	1038·9	1039·3
30·7	1039·6	1040·0	1040·3	1040·6	1041·0	1041·3	1041·7	1042·0	1042·3	1042·7
30·8	1043·0	1043·3	1043·7	1044·0	1044·4	1044·7	1045·0	1045·4	1045·7	1046·1
30·9	1046·4	1046·7	1047·1	1047·4	1047·7	1048·1	1048·4	1048·8	1049·1	1049·4

					Thousandths of an inch					
Inch ..	..	·001	·002	·003	·004	·005	·006	·007	·008	·009
Millibars	..	·0	·1	·1	·1	·2	·2	·2	·3	·3

Appendix 3

TABLE FOR CONVERTING DEGREES FAHRENHEIT (F) INTO DEGREES CELSIUS (C)

F	C	F	C	F	C	F	C	F	C	F	C
00	−17·8	20	−6·7	40	4·4	60	15·6	80	26·7	100	37·8
00·5	−17·5	20·5	−6·4	40·5	4·7	60·5	15·8	80·5	26·9	100·5	38·1
01	−17·2	21	−6·1	41	5·0	61	16·1	81	27·2	101	38·3
01·5	−16·9	21·5	−5·8	41·5	5·3	61·5	16·4	81·5	27·5	101·5	38·6
02	−16·7	22	−5·6	42	5·6	62	16·7	82	27·8	102	38·9
02·5	−16·4	22·5	−5·3	42·5	5·8	62·5	16·9	82·5	28·1	102·5	39·2
03	−16·1	23	−5·0	43	6·1	63	17·2	83	28·3	103	39·4
03·5	−15·8	23·5	−4·7	43·5	6·4	63·5	17·5	83·5	28·6	103·5	39·7
04	−15·6	24	−4·4	44	6·7	64	17·8	84	28·9	104	40·0
04·5	−15·3	24·5	−4·2	44·5	6·9	64·5	18·1	84·5	29·2	104·5	40·3
05	−15·0	25	−3·9	45	7·2	65	18·3	85	29·4	105	40·6
05·5	−14·7	25·5	−3·6	45·5	7·5	65·5	18·6	85·5	29·7	105·5	40·8
06	−14·4	26	−3·3	46	7·8	66	18·9	86	30·0	106	41·1
06·5	−14·2	26·5	−3·1	46·5	8·1	66·5	19·2	86·5	30·3	106·5	41·4
07	−13·9	27	−2·8	47	8·3	67	19·4	87	30·6	107	41·7
07·5	−13·6	27·5	−2·5	47·5	8·6	67·5	19·7	87·5	30·8	107·5	41·9
08	−13·3	28	−2·2	48	8·9	68	20·0	88	31·1	108	42·2
08·5	−13·1	28·5	−1·9	48·5	9·2	68·5	20·3	88·5	31·4	108·5	42·5
09	−12·8	29	−1·7	49	9·4	69	20·6	89	31·7	109	42·8
09·5	−12·5	29·5	−1·4	49·5	9·7	69·5	20·8	89·5	31·9	109·5	43·1
10	−12·2	30	−1·1	50	10·0	70	21·1	90	32·2	110	43·3
10·5	−11·9	30·5	−0·8	50·5	10·3	70·5	21·4	90·5	32·5	110·5	43·6
11	−11·7	31	−0·6	51	10·6	71	21·7	91	32·8	111	43·9
11·5	−11·4	31·5	−0·3	51·5	10·8	71·5	21·9	91·5	33·1	111·5	44·2
12	−11·1	32	0·0	52	11·1	72	22·2	92	33·3	112	44·4
12·5	−10·8	32·5	+0·3	52·5	11·4	72·5	22·5	92·5	33·6	112·5	44·7
13	−10·6	33	+0·6	53	11·7	73	22·8	93	33·9	113	45·0
13·5	−10·3	33·5	0·8	53·5	11·9	73·5	23·1	93·5	34·2	113·5	45·3
14	−10·0	34	1·1	54	12·2	74	23·3	94	34·4	114	45·6
14·5	− 9·7	34·5	1·4	54·5	12·5	74·5	23·6	94·5	34·7	114·5	45·8
15	− 9·4	35	1·7	55	12·8	75	23·9	95	35·0	115	46·1
15·5	− 9·2	35·5	1·9	55·5	13·1	75·5	24·2	95·5	35·3	115·5	46·4
16	− 8·9	36	2·2	56	13·3	76	24·4	96	35·6	116	46·7
16·5	− 8·6	36·5	2·5	56·5	13·6	76·5	24·7	96·5	35·8	116·5	46·9
17	− 8·3	37	2·8	57	13·9	77	25·0	97	36·1	117	47·2
17·5	− 8·1	37·5	3·1	57·5	14·2	77·5	25·3	97·5	36·4	117·5	47·5
18	− 7·8	38	3·3	58	14·4	78	25·6	98	36·7	118	47·8
18·5	− 7·5	38·5	3·6	58·5	14·7	78·5	25·8	98·5	36·9	118·5	48·1
19	− 7·2	39	3·9	59	15·0	79	26·1	99	37·2	119	48·3
19·5	− 6·9	39·5	4·2	59·5	15·3	79·5	26·4	99·5	37·5	119·5	48·6

Appendix 4

RADIO-SONDE

A radio-sonde is a balloon-borne instrument about the size of a transistor radio which is used for 'sounding' the atmosphere. Radio-sondes are launched twice daily (0001 and 1200 GMT) at several stations in each country throughout the world and by the few weather ships on station on the oceans. The balloon, roughly 2 m in diameter, rises at roughly 300 m/min and generally reaches a height in excess of 30 km before it bursts. A parachute is then activated to control the rate of descent in the interests of safety.

The instrument carries three sensing elements, for temperature, humidity and pressure, whose readings are alternately switched into a small transmitter. A light radar reflector is attached to the balloon so that the launching station, equipped with radar as well as a radio receiver, can obtain winds by radar tracking as well as values of temperature, humidity and pressure throughout the 'flight'.

The results are plotted on aerological diagrams as a type of temperature (and humidity)/height graph. Various forms of these diagrams are used by different nations; in the UK a Tephigram is used (Diagrams 156, 157). Completed aerological diagrams such as the Tephigram indicate the stability and instability of the atmosphere (Chapter 5) in the vicinity of the radio-sonde station and fairly readily show what cloud will occur and how deep this cloud will be. They also provide information for upper air charts.

Pressure decreases logarithmically with increasing height, but in the same way as the surface pressure varies, so the pressure at any given height (the weight of the air column above that level) is also constantly varying. The pressures at a given height level, obtained from the network of radio-sonde stations, could be plotted on a chart for that height and isobars drawn to indicate the areas of upper highs and lows and to derive winds (from the isobar spacing and direction) to supplement the relatively few radar winds already.

In practice, a more convenient system, used internationally, is to plot the varying heights of a given pressure level and to draw contours for that pressure level. Geostrophic relationships are applied to derive the winds from the contours in the same way as with isobars. The levels normally plotted are 850 mb (approx 1,800m), 700 mb (approx 3,000m), 500 mb (5,600m), 300 mb (9,200 m), 250 mb (10,300 m), 200 mb (11,800m), 150 mb (13,600 m) and 100 mb (16,200m). An example of a plotted and analysed 300 mb chart is given in Diagram 158.

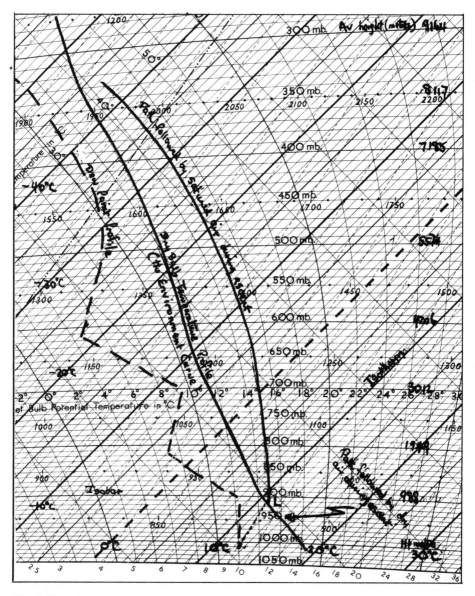

156. Radio-sonde ascent plotted on a Tephigram – unstable air.

The Tephigram* can be treated as a temperature-pressure (height) graph. The Y axis is pressure, decreasing upwards (the lines – isobars – are slightly curved and are labelled 100 mb, 950 mb, 900 mb etc). The average heights in metres of the 'main' pressure levels are given on the right.

The X axis (skew) is temperature, increasing towards the bottom right and its lines slope upwards to the right.

The solid plotted line is the obtained dry-bulb temperature profile while the dark broken line is the dew-point profile. These are the 'environment curves' – the results of the sounding.

Air which is forced to rise either through convection (surface heating) or orographically (over mountains) will cool at the rate shown by the straight lines sloping upwards to the left, *while it remains dry*. (This rate of cooling is called the Dry Adiabatic Lapse Rate.) The water vapour in the air rising from the surface remains unchanged in amount so that as the air cools, the relative humidity increases.

Saturation (condensation) occurs at the level where the surface dew-point is projected along the faint pecked lines to meet the projected dry-bulb line (at point L). After condensation, the path followed by rising air follows lines curving upwards towards the left. The new rate of cooling is less than the earlier (dry) rate because latent heat is released during condensation and this reduces the rate of cooling. (This reduced rate is called the Saturated Adiabatic Lapse Rate.)

If the temperature of the rising air

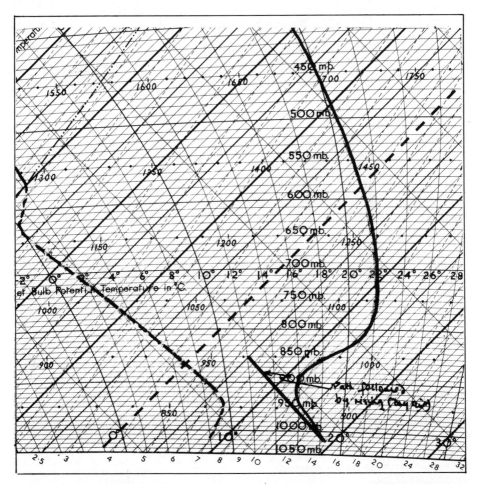

157. Radio-sonde ascent plotted on a Tephigram – stable air.

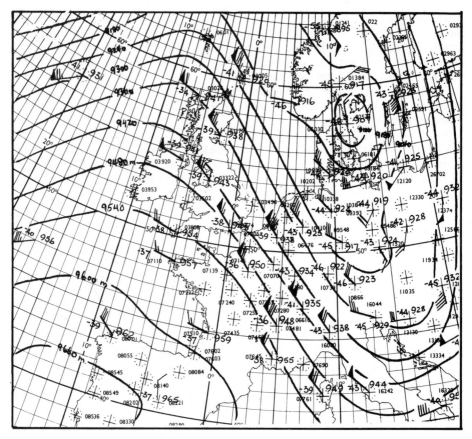

158. Example of an upper air chart – 300 mb surface at 1200 GMT on 11 Aug 1979. Contours are drawn at 60 m intervals. The closer the contours the stronger the winds. Geostrophic relationships (scales) and Buys Ballot's Law apply on upper air charts.

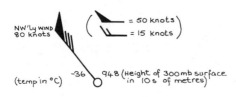

at any given level is warmer than the environment (i.e. it lies to the right of the environment curve) then the rising air, being less dense than its surround-

ings, will continue to ascend. E.g. at 600 mb level the rising air temperature is minus 8°C while the environment temperature (that of the air already at that level) is minus 13.5°C; the rising air is 5.5°C warmer than its environment and it will continue to rise. This process continues until the temperature of the rising air is no longer warmer than its environment.

In Diagram 156 (unstable air) condensation occurs in the rising air at about 910 mb (about 980 metres) and continues until the 310 mb level (about 9000 metres) is reached. Throughout this height range there would be convective cloud (cumulonimbus) producing heavy showers and thunderstorms.

The example of Diagram 157 indicates a Stable Air mass. In this case surface air which is forced to rise is colder at every level than the environment and so (being more dense) it would return to the surface before condensation can occur. There would be no cloud associated with this example.

* The proper (rectangular) axes of the Tephigram are Temperature (T) and Entropy (ϕ); the latter is a mathematical concept which will receive no further mention here.

Appendix 5

SATELLITES

Meteorological satellites have greatly increased the amount of weather information available at the main analysis centres in each country. This is particularly so over the oceans and deserts where there are few formal weather observations and in many areas none at all. The first such satellite launched in 1960 immediately confirmed its suitability as an observation platform. At first coverage was restricted to tropical regions, mainly to locate tropical revolving storms, and then was only available to a few selected ground receiving stations.

In 1964 a new generation of satellites was introduced. These were placed into almost circular polar orbits at heights which varied from satellite to satellite between about 650 and 1500 km. They used automatic picture transmission (APT) which could be intercepted by any station equipped with the appropriate, and fairly inexpensive,

receiving and recording equipment. Their current successors orbit at heights of about 850 km which gives an orbit time of about 100 minutes. A sketch of a polar orbiting satellite track is given in Diagram 159.

The orbit is deliberately chosen so that the satellite passes over each place at almost the same time each day. This comes about because the satellite orbit is fixed in space and its plane therefore remains at a constant, small angle to the direction of the sun (seen from the earth's centre). The satellite therefore keeps pace with the apparent 'westward moving' sun, i.e. it passes over each place at almost the same local time each day. To ensure it passes all places irrespective of latitude at the same local time, the satellite orbit is not truly polar. It is inclined to the meridians at a sufficient angle to make enough westing as it moves polewards to keep in step with the sun. An angle between the satellite north-bound track

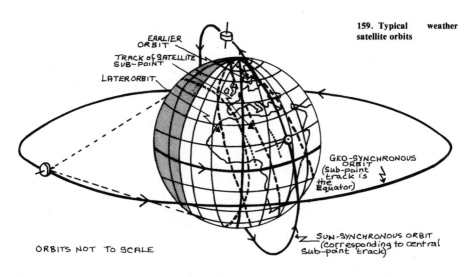

159. Typical weather satellite orbits

EARLIER ORBIT

TRACK of SATELLITE SUB-POINT

LATER ORBIT

GEO-SYNCHRONOUS ORBIT (Sub-point track is the Equator)

SUN-SYNCHRONOUS ORBIT (corresponding to central Sub-point track)

ORBITS NOT TO SCALE

222

and the equator of about 80° meets this requirement. The local times chosen for satellite pass are normally around 0900 or 1500 to ensure sufficient shadow effect, a useful aid to analysis. These polar orbiting satellites are said to be in sun-synchronous orbit.

Of course when a satellite passes over a given place north-bound during the day it will also pass south-bound during the night. For some time satellites have carried infra-red equipment so that useful cloud pictures are also available at night.

Another, different, generation of satellites was introduced in 1966. These are placed in orbit over the equator at a height of about 35,900 kilometres which ensures an orbit time of 24 hours. They are in geo-synchronous or geo-stationary orbit, i.e. they constantly remain over the same point on the equator. These satellites transmit pictures each 30 minutes covering a circular area ranging from about 50°N to 50°S, and 50° east and west along the equator, from the satellite sub-point (see Diagram 159).

Earlier satellites transmitted their pictures by television but these have been superseded by sensors which separately detect radiations in both the visual wavelengths (reflected sunlight) and in the infra-red wavelengths (heat radiation from land, ocean or cloud-top, giving a measure of temperature). While a polar-orbiting satellite tracks along a 'near-meridan' of longitude its sensors are continuously scanning east to west to cover a swath about 3,500 km wide. The north-south range at a given receiving station varies from about 30° to 60° of latitude according to whether the orbit is overhead or only a little above the horizon.

In the visual range the signals are finally reproduced as 'grey-scale' pictures, as though by a camera using black and white film: areas of low reflectivity such as forest or grassland and the oceans appear darker shades of grey, whereas areas of high reflectivity such as clouds, snow or ice and desert appear lighter grey or white. Thick clouds (great depth) and sea fog (very high moisture content) are strongly reflective and appear very white. Infra-red signals are also coded into grey-scale pictures. The convention used is that cold areas such as high cloud, snow and ice appear white whereas warm areas appear black, with numerous grey shades in between.

These pictures from both polar and equatorial orbiting satellites reveal patterns of cloud which are fairly easily identifiable as depressions and their fronts, anticyclones (often cloud-free) and areas of thunderstoms (sometimes organized into convergence zones or line-squalls). The type of cloud may be identified, especially when analysing simultaneous visual and infra-red pictures, and patterns within cloud sheets sometimes indicate the wind direction at cloud level. Equatorial orbiting satellites (geo-stationary) are particularly useful here since wind speed and direction may be obtained by examining consecutive pictures at about half-hourly intervals. Both kinds of satellite are invaluable for the surveillance of tropical revolving storms.

Examples of satellite pictures are given in the Plates section.

Index

Other Stanford Maritime books on navigation, seamanship and cruising

Navigation for Yachtsmen *Mary Blewitt*
Coastwise Navigation *Gordon Watkins*
Exercises in Coastal Navigation *G. W. White*
Celestial Navigation for Yachtsmen *Mary Blewitt*
Exercises in Astro-Navigation *Gordon Watkins*
Basic Principles of Marine Navigation *D. A. Moore*
Marine Chartwork 2ND ED. *D. A. Moore*
Race Navigation* *Stuart Quarrie*
Traverse and Other Tables
Burton's Nautical Tables
Planispheres
Stars at a Glance
Outlook: Weather Maps and Elementary Forecasting *G. W. White*
International Light, Shape and Sound Signals *D. A. Moore*
Guide to the Collision Avoidance Rules *Cockcroft & Lameijer*
Stanford's Sailing Companion 4TH ED. *R. J. F. Riley*
Stanford's Tidal Atlases *Michael Reeve-Fowkes*
Norwegian Cruising Guide *Mark Brackenbury*
Baltic Pilot* *Mark Brackenbury*
Frisian Pilot – Den Helder to the Kiel Canal *Mark Brackenbury*
Barge Country – the Netherlands Waterways *John Liley*
France – the Quiet Way *John Liley*
Brittany and Channel Islands Cruising Guide *David Jefferson*
Scottish West Coast Pilot *Mark Brackenbury*
Shell Encyclopaedia of Sailing *Michael Richey*
Practical Yacht Handling *Eric Tabarly*
Motorboat & Yachting Manual 19TH ED. *Dick Hewitt*
Better Boat Handling* *Des Sleightholme*
After 50,000 Miles *Hal Roth*
Capsize in a Trimaran *Nicolas Angel*
Two Against Cape Horn *Hal Roth*
Blown Away *Herb Payson*
Seven Times around the Sun* *Nicole Nealey-van der Kerchove*
A Thirst for the Sea: the Sailing Adventures of Erskine Childers
  Hugh Popham

* In preparation at the time of publication of this book

For a complete list of nautical books and charts, write to the Sales Manager,
Stanford Maritime Ltd, 12–14 Long Acre, London WC2E 9LP